D1215352

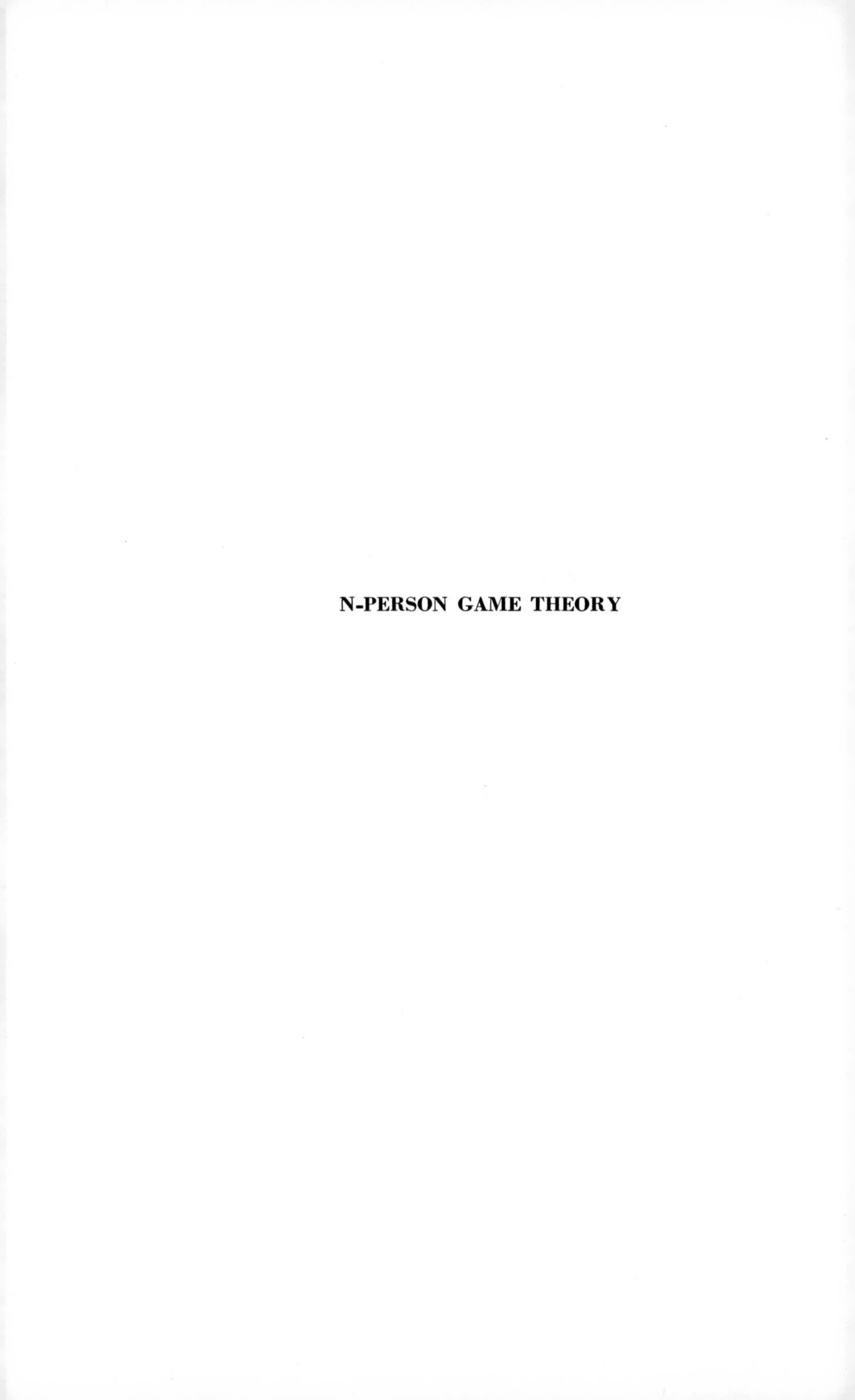

N-PERSON GAME THEORY

Ann Arbor
Science
Library

N-Person Game Theory

Concepts and Applications

by Anatol Rapoport

ANN ARBOR
THE UNIVERSITY OF MICHIGAN PRESS

Copyright © by The University of Michigan 1970
All rights reserved
SBN 472-00117-5
Library of Congress Catalog Card No. 79-83451
Published in the United States of America by
The University of Michigan Press and simultaneously
in Don Mills, Canada, by Longmans Canada Limited
Manufactured in the United States of America

Preface

This book is a sequel to *Two-Person Game Theory: The Essential Ideas* (University of Michigan Press, 1966). It is addressed to the same audience: to people with little mathematical background but with an appetite for rigorous analysis of the purely logical structure of strategic conflict situations.

In the preface to *Two-Person Game Theory* I explained why I found it necessary to separate the expositions of Two-person and N-person game theory. The former can be presented with a minimum of (mostly familiar) mathematical notation; the latter cannot. I remain convinced that unfamiliar mathematical notation scares at least as many people away from mathematical treatments of important subjects as the difficulty of following mathematical reasoning. The situation in the study of Russian is somewhat similar but with an important difference. Many people think that Russian is difficult to learn because it is written in an unfamiliar alphabet. Russian is, to be sure, comparatively difficult for non-Slavic speakers, but certainly not because of its non-Latin alphabet. The Cyrillic alphabet can be learned in an hour. The difficulties stem largely from the fact that Russian is a more inflected language than the modern Germanic and Romance languages, so that there is more grammar to learn. I wish I could say the same for mathematics: that the difficulties of notation are trivial and that only the difficulties of mathematical "grammar" (mode of reasoning) need to be overcome. Unfortunately, this is not the case. Mathematical notation and mathematical reasoning are much more intertwined than alphabet and grammar;

so that one cannot really learn to read mathematical notation without acquiring a certain degree of mathematical maturity.

The interdependence between mathematical notation and mathematical logic suggests the task of mathematical pedagogy: one must constantly emphasize the essential connections between the symbols and the concepts for which they stand. This emphasis is particularly important in set theory, one of the mathematical pillars on which N-person game theory rests. When the reader has learned to associate quickly the concepts with their representations, he is well on the way toward understanding set-theoretic reasoning and has mastered one half of the conceptual repertoire that underlies game theory. The other half is the notion of multi-dimensional space as the set of all possible n-tuples of numbers. In Two-person game theory, this notion presents no difficulty. There being only two players, all possible payoffs of a game are pairs of numbers, representable on two-dimensional diagrams. If $n = 3$, we can still resort to projections of three-dimensional figures. For $n > 3$, visual intuition fails. One must learn to think in terms of visually unrepresentable "spaces." Here again, once one has learned to "read" properly, the conceptual difficulties begin to resolve themselves quite rapidly.

The Introduction is a summary of mathematical concepts that I believe to be sufficient for understanding the essential ideas of N-person game theory. The ideas themselves (mostly in a purely logical context) are presented in Part I. "Applications" are discussed in Part II. The Introduction to Part II will hopefully forestall misunderstanding concerning the meaning of "applications" in the context of game theory. More will be said on this matter in the last two chapters.

The scope of the book covers the essential ideas developed in the original formulation of N-person game theory by Von Neumann and Morgenstern and the subsequent extensions by the present generation of game theoreticians. In their book *Games and Decisions*, Luce and Raiffa have already covered practically all of the significant advances up to 1957. Since then, two more volumes of *Contributions to the Theory of Games* (Annals of Mathematics Studies series, Princeton University Press) have appeared, as well as many separate journal articles, some proceedings of conferences on game theory, and numerous memoranda and preprints. These were my main sources.

The reader will note that the authors cited are predominantly American and Israeli. This reflects the continued interest in the United States and in Israel in the application potential of game theoretic ideas to social science. There is also a large Russian literature; but, to the extent that I have examined it, it is of interest only to the mathematical specialist, and so falls outside the scope of this book.

I take pleasure in thanking the University of Michigan Press for continued encouragement. I am indebted to Professors R. M. Thrall and to William F. Lucas for their critical reading of the manuscript and for many helpful suggestions. My heartfelt gratitude goes once more to Claire Adler, who has given invaluable editorial assistance, and to Dorothy Williams Malan for help in the preparation of the manuscript.

Contents

Introduction

Some Mathematical Tools

Game theory is properly a branch of mathematics. As such it is concerned with assertions which can be proved to be true if certain other assertions are true. This way of establishing truth by reference to assertions previously established as true would lead to infinite regress unless certain fundamental assertions were simply accepted as true (without proof). These basic assertions are called *axioms.* Assertions whose truth is derived by logical proof are called *theorems.* Game theory, like any other mathematical theory, is essentially a collection of theorems derived from axioms.

The terms (words and phrases) of a mathematical proposition must be precisely defined. The definitions contain other words, which must also be defined. Definitions likewise would lead to infinite regress or to circularity unless some terms were simply accepted as understood. These fundamental accepted terms are called *primitive terms.* Every mathematical theory must contain some primitive terms. All other terms must be defined by reference to these.

It is important to keep in mind that a mathematical concept is never defined "approximately," i.e., with a tacit assumption that its meaning is intuitively clear. A mathematical term is always defined exactly, so that there can be no dispute about its meaning. Similarly a mathematical theorem is never approximately or "reasonably" true, or true "with a high degree of probability." It is always absolutely true (assuming the axioms to be true). To assert that a theorem is false means to deny one or more of the axioms. However, to assert that a theorem is true does not necessarily mean to assert the truth of all the axioms. Some theorems remain true even if some of the axioms of a mathematical theory are rejected.

To accept the truth of the axioms is simply to agree to assume them to be true. However, the consistency of the axioms with each other is sometimes in question. Then, if any two or more axioms of an alleged mathematical theory are found to be inconsistent with each other, the whole theory collapses.

The truth of a mathematical assertion does not depend on observations. In this way, pure mathematics differs sharply from the natural sciences, where truth must be in conformity with observations. This dependence of the truth of assertions on observations (which are always subject to error) and also on induction (deriving general propositions from observations of special cases) necessitates the admission into the natural sciences of assertions which are approximately or reasonably true, or true with a high degree of probability.

Just as the natural sciences have developed complex and intricate methods of *observation* (microscopes, telescopes, laboratory tests, etc.), so mathematicians have developed complex and intricate methods of *deduction.* These latter methods are essentially rules for handling symbols. Mathematical symbols are powerful mnemonic devices which enable mathematicians to make complex deductions without getting entangled in words.

As an example, try to prove the following theorem of arithmetic without the use of symbols. "If a certain number is decreased by one and increased by one, and the two resulting numbers are multiplied, the product is equal to the square of the original number decreased by one." It can be done, but it is not easy. Mathematical symbolism makes the proof simple. We designate "any number" by x, the two numbers to be multiplied by

$x + 1$ and $x - 1$, and perform the multiplications by the well known rules of algebra (which are simply rules for handling symbols). We get

$$(x + 1)(x - 1) = x(x - 1) + (x - 1) \\ = x^2 - x + x - 1 = x^2 - 1, \quad (0.1)$$

which is what we had to prove.

Higher branches of mathematics (e.g., the calculus, differential equations) use special symbols which facilitate the deduction of the theorems in those domains, namely theorems concerning derivatives, limits, integrals, etc. All these concepts are symbolized, and the mathematical theories which use these concepts are essentially based on techniques of manipulating these symbols in chains of logical deduction.

Sets

The branches of mathematics on which game theory (especially N-person game theory) leans heavily is called the *theory of sets.* Unfortunately this branch of mathematics has not until recently been taught in conventional mathematical curricula. The situation has changed since the introduction of so-called New Math, of which set theory is an important sector. In general, this change in the mathematical curriculum has been an outstanding success. Possibly mathematics has always been a fascinating and exciting subject for active young minds; and it remains only to develop an approach which would effectively illuminate the logical interrelations not only between mathematical propositions but also between branches of mathematics (arithmetic, algebra, and geometry), to allow the youngsters to "stay" with the subject as it develops, instead of getting lost in rules and drills.

If any branch of mathematics can be said to be the foundation of all the others, it is certainly set theory. It turns out that acquaintance with set theory is also indispensable to acquire a feeling for the mathematical mode of reasoning, to understand probability theory, decision theory, and the theory of games (especially the portions related to games with more than two players). Elements of set theory are now included in practically every elementary textbook on probability theory. Elementary textbooks on decision theory and on N-person game theory are still scarce; and so, for the convenience of the reader, an explanation of the essential ideas of set theory are given here, together with some other mathematical ideas needed in the exposition of N-person game theory.

The fundamental concepts of set theory are "an element," "a set," and "belonging to a set." Although formally these concepts must remain undefined (they are the primitive terms of set theory), it is easy to illustrate them in terms of common experience. An *element* can be thought of as anything that must remain itself and is not further analyzed (into constituent parts) in the course of a discussion. A *set* is a collection of such things, which remains fixed throughout a discussion. An element "*belongs*" to a set if it is a member of this collection. The assumption is that, once a set is specified, it is always possible to decide whether a given object is or is not an element of the set. Better said, a set is specified *if* it is possible to decide unambiguously whether something does or does not belong to it; but not otherwise. If i is an element and N a set, read $i \in N$: "i belongs to N," and $i \notin N$: "i does not belong to N."

To take an example, consider the expression "the countries of Europe." The set is well defined if everyone who uses this expression agrees on what is or is not a country of Europe. Arguments might arise about whether Scotland or East Germany are countries or parts of countries; whether San Marino and Andorra are countries; whether USSR, Turkey, and Denmark are countries of Europe (in view of the fact that most of their territories are outside of Europe). Until and unless such arguments are settled one way or another, the "countries of Europe" is not defined as a set.

To avoid fruitless arguments, "the countries of Europe" could be designated simply by enumerating the countries. Since there are fewer than thirty of them, this is easy to do. For some sets this is impractical; for example, "all male human beings in Michigan." This is a set if and only if the recognition of "something" as "a male human being in Michigan" is definitive. It is so in the overwhelming majority of cases, although conceivably difficulties might arise with hermaphrodites or with persons who straddle a state border. These may or may not become practical difficulties. At any rate, the criterion for specifying a set is clear, and it is intimately bound up with the criteria for specifying an element of the set and the notion of "belonging to a set."

In mathematics, many sets contain infinite numbers of elements, for example, "all even integers" or "all the rational numbers between zero and one inclusive." In specifying such infinite sets it is always assumed that a procedure is available for deciding whether a given object is or is not an element (member) of the set.

Sets are mathematical objects, like numbers; and like numbers, they can be "operated on" by mathematical procedures. Operations can be performed on single sets, pairs of sets, triples, etc., as in the case of numbers. For example, the following operations can be performed on single numbers: (a) squaring the number; (b) taking the square root of the number; (c) reversing the sign; (d) taking the logarithm; etc. The result of such an operation may be once again a number, or it may be a set of numbers. For instance, squaring -2 gives $+4$; taking the square root of 4 yields two numbers, $+2$ and -2; taking the logarithm yields an infinity of numbers.[1]

Among the operations on pairs of numbers are the fundamental operations of arithmetic (addition, multiplication, subtraction, and division) and some others, e.g., raising one number to a power designated by another, finding the least common multiple of two integers, etc.

In set theory an important operation on a single set is taking the *complement* of the set. The complement can be taken only with respect to a set which includes the given set (a *superset*). This operation determines a set whose members are all those members of the superset which are not members of the set operated upon. For example, "all human beings in Michigan" is a superset which includes the set "all male human beings in Michigan." Then the complement of the latter set with respect to the superset would be "all female human beings in Michigan" (assuming that the sex of any human being is unambiguously established). On the other hand, if we take "all mammals in North America" as a superset, then the complement of "all male human beings in Michigan" would be "all mammals in North America except the male human beings in Michigan." Note that the elements of a set together with its complement with respect to a superset constitute the superset.

We shall now introduce mathematical notation. All elements will be designated by lower case roman letters, and sets of these elements by capital roman letters. Thus in our example, a, b, c, . . . , may represent individual human beings, M the set of males, F the set of females, H the set of all human beings. The operation of taking the complement will be designated by the minus sign. Thus, $-\mathrm{M}$ stands for "the complement of M" (assuming that the superset with respect to which the complement is taken has already been designated). We can thus write

$$-\mathrm{M} = \mathrm{F}, \tag{0.2}$$

analogous to the way we write $2^2 = 4$ in arithmetic.

If the superset has not been specified, it must be specified in the notation of the complement. Thus, to indicate that H ("all human beings in Michigan") is the superset, we must write for the complement of M

$$\mathrm{H} - \mathrm{M} = \mathrm{F}. \tag{0.3}$$

This notation suggests subtraction. In fact, taking the complement can be viewed as a special case of *subtraction,* an operation on pairs of sets. Thus, if S and T are sets, $S - T$ is a set whose members are those members of S which are not members of T.

Consider now the set $S - S$. By definition of subtraction (of sets), this set has no members. It is called the *null* (or *empty*) set and is designated by $\emptyset$, a symbol which resembles 0 but is not identical with it, since 0 is a symbol for a number, not a set. Note that if S includes T, then $T - S = \emptyset$; if no member of S is a member of T, then $T - S = T$. If S and T have no members in common, we call them *disjoint.*

We have used the equality sign to denote the identity of sets, implying that two sets are identical if every member of one is also a member of the other, and vice versa. Equation (0.2) reads in English "Every human being in Michigan who is not a male is a female, and every human being in Michigan who is not a female is a male."

If one set includes another (i.e., if every member of the latter is a member of the former), we denote the relation by $\supseteq$. Thus we can write

$$\mathrm{H} \supseteq \mathrm{M} \tag{0.4}$$

(H includes M) to say "Every human male in Michigan is a human being in Michigan." We can also write

$$\mathrm{M} \subseteq \mathrm{H} \tag{0.5}$$

(M is included in H) to say exactly the same thing.

Clearly, if two sets include each other, they are identical. Thus, if $S \supseteq T$ and $S \subseteq T$, then $S = T$. However, if $S \supseteq T$ but $S \nsubseteq T$ (a vertical line through a symbol denoting a relation means that the relation does not obtain), then we write $S \supset T$, and say that T is a *proper subset* of S. If $S \supseteq T$, we can still say that T is a *subset* of S, but not necessarily a proper subset.

The basic operations on pairs of sets are the *union* and the *intersection.* These operations are meaningful without reference to a superset. The union of two sets is a set whose elements belong *either* to the one *or* to the other, *or* to both. The intersection of two sets is a set whose elements belong to *both* sets.

For example, let A be the set of adults in Michigan (assumed specified). Then

$$A \cup M \tag{0.6}$$

(the union of A and M) denotes all persons who are either adult or male or both (roughly all men, women, and boys). The intersection of A and M is denoted by

$$A \cap M. \tag{0.7}$$

It denotes all men. Clearly

$$A \cap M \subseteq A \subseteq A \cup M \tag{0.8}$$

and

$$A \cup M \supseteq M \supseteq A \cap M; \tag{0.9}$$

that is to say, the union of two sets includes either of them, and any set includes the intersection of itself with another. It follows that every set includes $\emptyset$, and $\emptyset$ is included in every set.

If one set is included in another, then the intersection of the two is the included set and the union of the two is the including set. Thus

$$\begin{aligned} A \cap H = A; \quad M \cap H = M \\ A \cup H = H; \quad M \cup H = H. \end{aligned} \tag{0.10}$$

The obviousness of such conclusions may give the impression that set theory is no more than a formalization of trivialities. That this is not the case will, I hope, become apparent as we develop further consequences of our simple rules of operation.

The reader may easily verify the following equalities:

$$(A \cup B) \cup C = A \cup (B \cup C) \qquad (0.11)$$

$$A \cup B = B \cup A \qquad (0.12)$$

$$(A \cap B) \cap C = A \cap (B \cap C) \qquad (0.13)$$

$$A \cap B = B \cap A. \qquad (0.14)$$

The following equalities are, perhaps, somewhat less obvious:

$$A \cap (B \cup C) = (A \cap B) \cup (A \cap C) \qquad (0.15)$$

$$A \cup (B \cap C) = (A \cup B) \cap (A \cup C). \qquad (0.16)$$

Let us illustrate the second. Let A represent all adults, B all males, C all persons born in Michigan. Then the left side of (0.16) represents all persons who are either adults or males born in Michigan. The right side represents all persons who are either adult or males and at the same time either adults or born in Michigan. The identity of the sets represented by the left and right sides of (0.16) is not immediately obvious, because ordinary language is too clumsy to describe the sets. We can clarify matters by a diagrammatic representation (the so-called *Venn* diagram) shown in Figure 1.

Operations of arithmetic and algebra strongly resemble those of set theory (but differ in some respects). We have, for example,

the commutative law of addition in algebra:

$$a + b = b + a, \qquad (0.17)$$

which corresponds to (0.12) above;

the commutative law of multiplication:

$$a \times b = b \times a, \qquad (0.18)$$

which corresponds to (0.14) above;

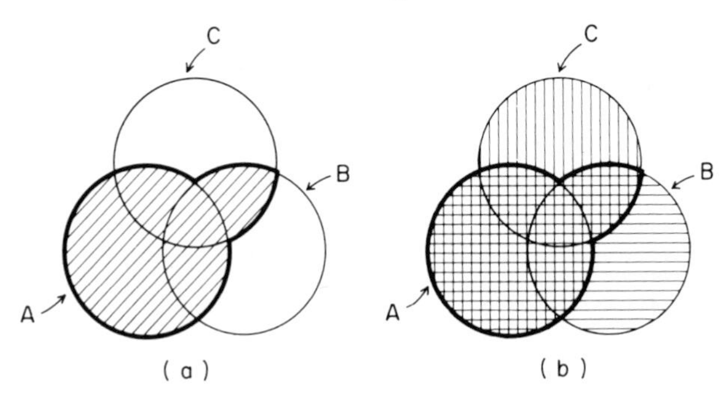

FIG. 1. In diagram (a) we have shaded the set A diagonally, then the intersection of B and C diagonally. The horizontally shaded area within the thick borders represents the union of these two sets. In diagram (b) we have shaded A and B horizontally, then A and C vertically. The area shaded both ways is the intersection. Note that both procedures yield the same set (included in the thick borders).

the distributive law of multiplication with respect to addition:

$$a \times (b + c) = (a \times b) + (a \times c), \tag{0.19}$$

which corresponds to (0.15) in sets.

Note, however, that in general

$$a + (b \times c) \neq (a + b) \times (a + c). \tag{0.20}$$

That is to say, there is no "distributive law" of addition with respect to multiplication in algebra; but there is in set theory [see equation (0.16)].

The operations of union, intersection, and complement can be combined to give the following equalities:

$$-(A \cup B) = (-A) \cap (-B). \tag{0.21}$$

$$-(A \cap B) = (-A) \cup (-B). \tag{0.22}$$

In English, the complement of the union of two sets is the intersection of their complements; the complement of the intersections of two sets is the union of their complements.

In what follows, all of the sets will be subsets of a fixed superset, I. As we have already seen, the union of any set S and its complement $-S$ with respect to I is I. The intersection $S \cap (-S)$ by definition comprises the elements which are members of both. But by definition of complement, there are no such elements. Therefore $S \cap (-S) = \emptyset$.

The set $\emptyset$ behaves like zero in many ways, not only because it is empty but also because of the way it enters into operations. Suppose that the operation $\cup$ is viewed as the analogue of "plus" in arithmetic, and $\cap$ as the analogue of "times." Observe that

$$A \cup \emptyset = A, \text{ like } a + 0 = a; \tag{0.23}$$

that is, $\emptyset$ behaves like 0 in "addition." Also,

$$A \cap \emptyset = \emptyset, \text{ like } a \cdot 0 = 0; \tag{0.24}$$

that is, $\emptyset$ behaves like 0 in "multiplication."

If $A \cap B = \emptyset$, A and B are clearly disjoint. A set and its complement are always disjoint; but, of course, there are many other sets disjoint with a given set, in particular all the subsets of its complement. In symbols, we can write

$$\text{If } A \cap B = \emptyset \text{ and } C \subseteq B, \text{ then } A \cap C = \emptyset. \tag{0.25}$$

The elements of a set, we have said, may be any "objects," where objects are not necessarily physical objects like tables, chairs, mice, or men, but simply any distinguishable entities, anything that can be designated and that remains clearly designated in the course of a discussion. Thus there is no logical difficulty in specifying a set comprising all telephone poles, all constellations, all angels, and all irrational numbers—provided we agree what these entities stand for, although it would be difficult offhand to say in what context this set would be of interest. In particular, elements of a set may themselves be sets.

If a set is defined by enumerating its elements, we can denote it by putting the elements in braces. Thus, the set of presidents of the United States can be designated by

$$P = \{w, a, j, m, \ldots, j^{(4)}\}, \qquad (0.26)$$

where w denotes Washington, a Adams, $j^{(4)}$ Johnson [written in 1967], etc.

Consider the set N comprising the elements a, b, c, d. We can write $N = \{a, b, c, d\}$. Consider now all of the *subsets* of N. They are:

$$\begin{aligned}&\emptyset; \{a\}, \{b\}, \{c\}, \{d\}, \{a, b\}, \{a, c\}, \{a, d\},\\ &\{b, c\}, \{b, d\}, \{c, d\}, \{a, b, c\}, \{a, b, d\}, \qquad (0.27)\\ &\{a, c, d\}, \{b, c, d\}, \{a, b, c, d\}.\end{aligned}$$

These subsets are identifiable objects and therefore can themselves be considered members of a set, namely the set of subsets of N. We shall denote this set of subsets and *its* subsets by script letters, in particular the whole set (of subsets of N) by $\mathfrak{N}$. For example, if $\mathfrak{Q}$ is the set of all the subsets of N containing two members of the original set N, then

$$\mathfrak{Q} = \{(a, b), (a, c), (a, d), (b, c), (b, d), (c, d)\}, \qquad (0.28)$$

where we have now put the subsets of N in parentheses instead of braces to indicate that they are now viewed as elements, not as sets.

The set $\mathfrak{Q}$ contains six elements (pairs of elements of N). The set $\mathfrak{N}$ contains 16 elements; the set of subsets of $\mathfrak{N}$ contains $2^{16} = 65{,}536$ elements.

N.B. The original *element* a is a member of N but not of $\mathfrak{N}$. The set {a}, consisting of the single element a, *is* an element of $\mathfrak{N}$, and we now can denote it by (a).

In the context of sets of subsets, we can define still another operation derived from the operation of complement but not identical with it. In the original set N, each subset had a complement with respect to N (as superset). For instance, the complement of {a, b} was {c, d}, etc. Now consider a subset of $\mathfrak{N}$, say

$$\mathcal{S} = \{(a, b), (a, c), (d)\}. \qquad (0.29)$$

Each of the elements in this subset has a complement with respect to N (not $\mathfrak{N}$); namely the complement of (a, b) is (c, d); the complement of (a, c) is (b, d); the complement of (d) is (a, b, c).

Now we define $\mathcal{S}^*$ (the *starred derivative* of $\mathcal{S}$) as a subset whose elements are complements of the elements of $\mathcal{S}$ considered as subsets of N. In our case

$$\mathcal{S}^* = \{(c, d), (b, d), (a, b, c)\}. \qquad (0.30)$$

Note that in this case the sets $\mathcal{S}$ and $\mathcal{S}^*$ have no common elements. Hence

$$\mathcal{S} \cap \mathcal{S}^* = \emptyset. \qquad (0.31)$$

However, a set and its starred derivative need not be disjoint. Consider the set

$$\mathfrak{J} = \{(a, b), (a, c), (c, d)\}. \qquad (0.32)$$

Then

$$\mathfrak{J}^* = \{(c, d), (b, d), (a, b)\}. \qquad (0.33)$$

It turns out that

$$\mathfrak{J} \cap \mathfrak{J}^* = \{(a, b), (c, d)\} \neq \emptyset. \qquad (0.34)$$

Clearly the operation (*) should not be confused with the operation (−). In our case, $-\mathfrak{J}$ contains 13 elements, i.e., all the elements of $\mathfrak{N}$ (i.e., subsets of N) except those in $\mathfrak{J}$.

In a way, the designation of the set of subsets of a set is a sort of expansion of the original set; at any rate, it is an operation performed on the original set. Unlike the previously described operations (e.g., "the complement of"), the operation of going to the set of subsets results in a set containing elements of *a different kind* from those of the original set. Recall that the elements of a set and those of its complement were elements of the same superset. But sets of elements are of a different logical order. They are not elements of the same superset. Thus no universal set can be logically constructed to contain all conceivable entities as elements.[2]

So far all the binary operations on sets resulted in sets whose elements were of the same kind, indeed members of one or both of the sets operated upon. We now introduce a binary operation on sets that results in a set with elements of a different logical order from those of the original sets. This operation is called the *cartesian product* and is denoted by $\times$.

Let A and B be sets. Then the cartesian product, $A \times B$, is defined as a set consisting of all *ordered pairs* of elements of which the first member is an element of the first set, and the second an element of the second set. Thus, if

$$A = \{a_1, a_2, \ldots, a_m\} \tag{0.35}$$

and

$$B = \{b_1, b_2, \ldots, b_n\} \tag{0.36}$$

then

$$\begin{aligned} A \times B = \{&(a_1, b_1), (a_1, b_2), (a_1, b_3), \ldots \\ &(a_2, b_1), (a_2, b_2), \ldots \quad \ldots \\ &\ldots (a_m, b_n)\}. \end{aligned} \tag{0.37}$$

Note that the *pair* (a_1, b_1) is not the same as the *set* comprising a_1 and b_1. In the latter, the order in which the elements are named does not matter. In the former, it does. We shall, however, at times include elements of sets, as well as of ordered n-tuples, in parentheses instead of braces. The meaning will be clear from the context.

Note also that the cartesian product operation is not commutative. In general, $A \times B \neq B \times A$. This is because in $B \times A$, the b's would be the first numbers of all the pairs.

For example, if A represents all men between the ages of 20 and 30 in New York, and B all women of the same age, then $A \times B$ contains all possible couples, named in the order (man, woman), in that age group. Neither the members of A nor those of B are members of $(A \times B)$. A member of $(A \times B)$ is a *couple*. Members of A and B are *individuals*.

It is possible to form a cartesian product of a set with itself. As in any other cartesian product, the members are ordered pairs of individuals, both members of the set. In some of these pairs, both members may be the same individual. There is no logical difficulty in conceiving such pairs. Imagine the set of outcomes of a toss of a coin. This set (call it S) contains two elements, h (heads) and t (tails). Then the set of outcomes of two consecutive tosses of the coin is a cartesian product of this set with itself, i.e., $(S \times S)$. It consists of the four pairs: $\{(h, h), (h, t), (t, h), (t, t)\}$.

The concept of cartesian product can be obviously extended to operations on three or more sets. The product then consists of elements which are ordered triplets, quadruplets, etc., of members of the original sets, one member from each set.

Operational Notation

A great deal of the difficulty experienced by non-mathematicians in reading discussions of mathematical subjects stems not so much from the complexity of the ideas as from a lack of experience in reading mathematical notation. Yet it is precisely the compactness of mathematical notation which enables the mathematician to use his rules of deduction without getting entangled in words. The reader will find it useful to acquaint himself with some notation which is all but indispensable in discussions of the mathematical theory of games.

The idea of a variable designated by a symbol, say x, is familiar to all with the barest minimum of mathematical literacy: x stands for *any* number which is a member of a given set of numbers. A simple extension is a variable with a subscript. Thus x_1 may stand for any payoff which player i, a member of the set $\{1, 2, \ldots, i, \ldots, n\}$ may get in a given game. The n-tuplet $(x_1, x_2, \ldots, x_i, \ldots, x_n)$ represents the sets of all such payoffs, one to each player. Further

$$\sum_{i=1}^{n} x_i = x_1 + x_2 + \cdots + x_i + \cdots + x_n, \quad (0.38)$$

i.e., the sum of the payoffs (if the payoffs are quantities which can be added). Suppose now S is a subset of N, the set of all players in a game. We wish to indicate the sum of the payoffs which accrue to the members of S. We write

$$\sum_{i \in S} x_i, \quad (0.39)$$

that is to say, only those x_i are to be added which accrue to players i who are members of S. Note that here i is a variable: it denotes any member of S. Similarly we can write

$$\sum_{i \in S \cup T} x_i \quad \text{or} \quad \sum_{i \in S \cap T} x_i \quad \text{or} \quad \sum_{i \notin S - T} x_i \quad (0.40)$$

to indicate respectively the sums of payoffs accruing to the players who belong to the set $S \cup T$, or $S \cap T$, or do not belong to $S - T$.

The symbol Π indicates product. Suppose x_i is the payoff to player i, and d_i a specified number associated with player i. Then $x_i - d_i$ is the player's payoff diminished by d_i, and

$$\prod_{i \in N} (x_i - d_i) \qquad (0.41)$$

denotes the product of all such numbers, one associated with each player who is a member of N.

The symbols Min and Max indicate respectively the smallest and the largest of a given set of numbers. Thus Min {1, 2, −3, 0} = −3; Max {1, 2, −3, 0} = 2.

Now let S = {1, 2, . . . , s}, T = {1, 2, . . . , t}. Then the cartesian product S × T = {(1, 1), (1, 2) . . . (2, 1), (2, 2) . . . (s, t) }. Each element of S × T can be designated by a symbol (i, j), (i = 1, 2, . . . , s; j = 1, 2, . . . , t), where the parentheses indicate the ranges of the variables i and j. Specifically, let S be the set of strategies available to player 1 in a Two-person game, and T the set of strategies[3] available to player 2. Note that now i and j designate *strategies*, not players. The symbol x_{ij} can then be used to designate the payoff to player 1 *if* he chooses strategy i *and* player 2 chooses strategy j. Now we can designate by

$$\operatorname*{Min}_{j} \{x_{ij}\} \qquad (0.42)$$

the smallest payoff which accrues to player 1 if he uses strategy i; that is, the smallest payoff which accrues to him (if he chooses strategy i) of all the payoffs which may result from player 2's choice of strategy. Note that this number no longer depends on j (since *all* the j's have been examined), but it still depends on i. $\operatorname*{Min}_{j} \{x_{ij}\}$ can therefore stand for a *set* of numbers with the subscript i, say $\{y_i\}$. We can now denote the largest of these numbers (as i ranges over player 1's choices of strategy) by $\operatorname*{Max}_{i} \{y_i\}$. The resulting number can also be denoted by

$$\operatorname*{Max}_{i} \operatorname*{Min}_{j} \{x_{ij}\}, \qquad (0.43)$$

which is to say, the largest (with respect to player 1's

strategies i) of the smallest (with respect to player 2's strategies j) of the payoffs which can accrue to player 1. This number is called the *maximin* payoff to player 1.

Relations

Of importance equal to that of the concept of set is the concept of *relation.* Formally a relation can be defined as a set. Let X and Y be two sets and $X \times Y$ their cartesian product. Then a relation R can be defined as a subset of $X \times Y$. The connection between this concept and the usual intuitive notion of "relation" is not obvious. Let us spell it out. Take the well known relation "greater than." Let X be the set of all real numbers. Then the cartesian product $X \times X$ is the set of all ordered pairs of real numbers. Now of any pair (x_1, x_2) it is either true or not true that $x_1 > x_2$. The subset of all pairs (x_1, x_2) for which it *is* true defines the relation "greater than."

Relations need not, of course, refer to magnitudes. Thus "father of," "prime minister to," "co-conspirator of" are all relations. Relations are classified by their properties. Some relations are *reflexive,* which means that every member of a set is in that relation to itself (as well as, possibly, to other members). For example, the relation "upon division by seven leaves the same remainder as" is a reflexive relation among integers. Every integer obviously stands in that relation to itself but also to other integers which differ from it by multiples of 7. On the other hand, the relation "is a square root of" is not reflexive, because not every number is a square root of itself.

Some relations are symmetric, which means that if x stands in relation R to y (written xRy), then also necessarily yRx. The relation "leaves the same remainder as" is symmetric, but clearly not the relation "is the assassin of."

Some relations are a-symmetric, which means that if xRy, then $y \not R x$. (We write a vertical bar through the symbol to denote that the relation does not obtain.) Thus the relation "greater than" is cleary a-symmetric since $x > y$ implies $y \not> x$. The relation "the husband of" is clearly a-symmetric.

Some relations are transitive, which means that if xRy and yRz, then necessarily xRz. "Greater than," "descendant of," "younger brother of," are all transitive relations.[4]

One might think that "a member of" is a transitive relation; but here one must be careful. An element may be a member of a set and the set a member of a set of sets, but the original element, considered as an element, is not a member of the set of sets.

Of particular interest are *complete* relations, those which obtain for any two elements in the set if they are properly ordered. Thus the relation "equal or greater than" is complete with respect to real numbers, since for any pair of such numbers (x, y) it is either true that $x \geqslant y$ or $y \geqslant x$ (possibly both, in which case $x = y$). If a complete relation is also transitive, it is possible to establish a *weak ordering* among the elements of the set. In the case of the relation "equal or greater than," the weak ordering would be in the order of magnitude, the order of equal elements being arbitrary. If a complete relation is also a-symmetric, it is possible to establish a *strong ordering*, where each element has a unique place in the order. For example, the relation "greater than" induces a strong ordering on the real numbers.

We shall have occasion to discuss a relation called *domination* among possible payoff disbursements of an N-person game. Because of its connotation of size (or strength), it is tempting to conclude that this relation induces at least a weak ordering among the payoff disbursements. This, however, is not so; the relation is not transitive, and this fact introduces a great many of the complexities which pervade N-person game theory.

Functions

A function is essentially a correspondence between members of one set and those of another, or sometimes within the same set. Many words or phrases in ordinary language denote such correspondences. As an example, consider the phrase "the mother of." Since every human being has or has had exactly one mother, we can in principle determine the particular woman who is (or was) the mother of any human being named. A correspondence is thus established between the set of all human beings on the one hand and the set of all women on the other. Note that whereas every mother is a woman, not every woman is a mother. Strictly speaking, then, the correspondence "the mother of" is established between the set of all human beings on the one hand and a subset of women on the other. In contrast, consider the correspondence "eldest son of" where the first set is the set of all parents and the second is the set of all eldest sons. Since not every parent has an eldest son, this correspondence is between a subset of parents and the entire set of eldest sons. When we speak of function, we always have in mind a correspondence between two sets, not necessarily distinct. To each member of the first there corresponds exactly one member of the second. The correspondence need not be one-to-one: the same member of the second set may correspond to more than one member of the first. The first set is called the *domain* of a function; the second its *range*. The member of the domain to which a member of the range corresponds is called the *argument* of the function; the corresponding member of the range is called the *value* of the function.

In elementary mathematics, functions are usually indicated by operations on numbers. Consider the function "the square of." The domain of this function may be taken as the set of all real numbers, since every real number has a square. On the other hand, if real numbers are taken as the domain, the range of the function will be only the positive real numbers and zero, since the square of every negative number is positive.

Functions will be denoted by letters followed by empty parentheses. Just as in algebra a letter may stand for any *number*, so a letter may designate any *function* which, recall, is not a number but a correspondence rule. For instance, f( ) may stand for the function "the cube of." Note that in the designation of the function, nothing follows "of." For this reason, nothing appears in the parentheses following f. We can extend the phrase "the cube of" by specifying what it is the cube of, say "the cube of seven." Note, however, that then the phrase "the cube of seven" no longer stands for a function: it stands for a number, specifically the value of the function whose argument is 7, i.e., 343. In functional notation this would be written $f(7)$. Similarly we can put into the parentheses any member of the domain of the function. When this is done, the entire expression will no longer stand for a function but rather for the *value* of the function corresponding to a given argument.

Conventionally, functions are often designated by expressions like $f(x)$, where x is supposed to mean any member of the domain. (We shall also do so on occasions.) There is some danger of confusion, since $f(x)$ also stands for the value of the function corresponding to the argument x. It is important to keep these two concepts distinct.

Functional notation is also frequently used to emphasize the dependence of one entity on another. For example, in Chapter 11, we shall say ". . . each syndicate S has at its disposal a threat strategy $\theta(S)$." The functional notation $\theta(S)$ is a reminder of the correspondence between syndicates and their respective threat strategies. In this notation, $\theta(-S)$ stands for the threat strategy at the disposal of the syndicate composed of the members of the complement of the set S.

Examples

Let f( ) stand for the correspondence which results when we add 7 to a number and square the result. Then the reader can verify

$$f(0) = 49; f(1) = 64; f(-1) = 36;$$

$$f(x) = x^2 + 14x + 49 \tag{0.44}$$

$$f(x - y) = x^2 + y^2 + 49 + 14x - 2xy - 14y. \tag{0.45}$$

It is important to get rid of the idea (sometimes implicit in elementary expositions) that every function can be expressed by a simple mathematical operation. To begin with, the objects among which functions establish correspondences need not even be mathematical, as we have seen in the example of humans-to-mothers. Secondly, the operations may change in the various subsets of the domains. For example, we may well define f( ) over the domain of real numbers as follows: When the argument is negative, f( ) stands for "the square of"; when the argument is 0, 23, or 14,539, the value of the function is zero; when the argument is positive (except the values mentioned), the value of the function is the logarithm of the number to base 10. It is difficult to imagine where such a function might arise, but it is a perfectly well defined one.

In game theory we shall have much occasion to discuss functions whose arguments are sets rather than numbers. At times the arguments may be sets of sets. The ranges of the functions may be numbers, or they may themselves be sets, or sets of sets.

For example, consider the sets which are the subsets

of the set $N = \{1, 2, 3\}$. Here 1, 2, 3 are merely names of elements (not numbers), such as the "names" of the three players of a Three-person game. The subsets are $\emptyset$, $\{1\}$, $\{2\}$, $\{3\}$, $\{1, 2\}$, $\{1, 3\}$, $\{2, 3\}$, $\{1, 2, 3\}$. Let now v( ) designate a function which assigns a particular number to each of these subsets of N. For example:

$$v\{\emptyset\} = 0; v(\{1\}) = -1; v(\{2\}) = -2;$$
$$v(\{3\}) = 0; v(\{1, 2\}) = 1; v(\{1, 3\}) = 1; \quad (0.46)$$
$$v(\{2, 3\}) = 4; v(\{1, 2, 3\}) = 5.$$

The domain of v( ) is the subsets of N. The range of v( ) is a subset of the real numbers. In game theory, this function usually has the following meaning: N designates a set of players of an N-person game. Subsets of this set can form coalitions. When these players play as coalitions, they can be sure of getting jointly at least a certain amount (positive or negative), which constitutes the *value* of the game to that coalition. Then the function v( ), called the *characteristic function* of the game, specifies the value of the game to each of all possible coalitions. Note that $v(S)$ is a number, the value of the game to coalition S; so is $v(-S)$, the value of the game to the coalition complementary to S. The expression

$$v(S) + v(-S) = v(N) \qquad (0.47)$$

says that whatever coalition S forms, the sum of the joint payoffs accruing to it and to the complementary coalition (if it too forms) is always the same number $v(N)$, namely the value of the game to the coalition which includes all the players. This will *not* always be the case. If it is the case, we shall say that the N-person game is *constant-sum*.

Both the domain and the range of a function may be sets, sets of sets, or even sets of sets of sets. Consider the following function ψ ( ):

$$\psi\{(1), (2), (3)\} = [\{(1, 2), (3)\}, \{(1, 3), (2)\}];$$
$$\psi\{(1, 2), (3)\} = [\{(1, 2, 3)\}, \{(1), (2), (3)\}]; \qquad (0.48)$$
$$\psi\{(1), (2, 3)\} = [\{(1, 2), (3)\}, \{(1, 3), (2)\}, \{(1, 2, 3)\}].$$

Note that parentheses enclose sets, which are themselves elements belonging to sets of sets enclosed by braces, which are themselves elements of sets of sets of sets enclosed by brackets.

We see that the argument of $\psi(\ )$ in each case is a certain set of subsets of N. These subsets are disjoint and exhaustive (i.e., include all the members of N). They can be viewed as *partitions* of N, e.g., particular ways in which the N players can break up into coalitions. The values of $\psi(\ )$ are *sets of partitions.* The function $\psi(\ )$ can be interpreted as follows: Given some ways in which the players have joined into coalitions, there are certain specified ways in which they can *re-partition* themselves. For example, if initially every player is in a coalition only with himself, i.e., the partition is $\{(1), (2), (3)\}$, then [to satisfy $\psi(\ )$] the only groupings that can occur are those in which player 1 joins with 2 or with 3: $[\{(1, 2), (3)\}, \{(1, 3), (2)\}]$. Next, if the partition is $\{(1, 2), (3)\}$, the only regroupings that can occur are those in which all three join together or the coalition of 1 and 2 breaks up: $[\{(1, 2, 3)\}, \{(1), (2), (3)\}]$. Finally, if the partition is player 1 vs. the coalition of 2 and 3, the following may happen: either 2 leaves 3 to join 1, or 3 leaves 2 to join 1, or all three may join together.

Note that the argument of $\psi(\ )$ is a partition, and the value (given the argument) is a set of partitions. We say $\psi(\ )$ is *on* partitions *to* sets of partitions.

Vectors, Spaces, and Simplexes

A vector is an ordered n-tuple of numbers; e.g., (3, 0, 4), (21, −3), (0, 0, 0, 0, 0), (1/2, 1/2, 1/2, 1/2), etc. The individual numbers are the *components* of the vector. The components may also be variables, e.g., (x, y, z), in which case we have a variable vector.

Vectors can be designated by single letters topped by an arrow. For example, $\vec{x}$ may stand for the n-tuplet $(x_1, x_2, \ldots, x_n)$, in which x_i is the i-th component of $\vec{x}$.

If two vectors have the same number of components, they can be "added." Thus, if $\vec{x} = (x_1, x_2, \ldots, x_n)$ and $\vec{y} = (y_1, y_2, \ldots, y_n)$, then the sum of the two vectors denoted by $\vec{z}$ is

$$\vec{z} = \vec{x} + \vec{y} = (x_1 + y_1, x_2 + y_2, \ldots, x_n + y_n). \quad (0.49)$$

In other words, the components of the sum of two vectors are sums of the corresponding components of the vectors summed.

Any vector can be multiplied by a number. Thus a vector $\vec{x}$ multiplied by a number c gives, as a product, the vector

$$c\vec{x} = (cx_1, cx_2, \ldots, cx_n). \quad (0.50)$$

In other words, in the product, every component is multiplied by c.

Addition of vectors having been defined, we can add $c\vec{x}$ and $(1 - c)\vec{y}$. In particular, if c is any number between 0 and 1, then $c\vec{x} + (1 - c)\vec{y}$ is the *weighted average* of vectors $\vec{x}$ and $\vec{y}$, where c and $1 - c$ are the weights. In the weighted average vector, each component is the same weighted average of the corresponding component of the two vectors.

Consider now all possible vectors (x, y) where x and y range over all real numbers. Each vector of this sort can be represented by a *point* in two-dimensional space (an infinite plane) with a coordinate system. Here x denotes the distance (positive or negative) from the vertical axis, i.e., the horizontal coordinate of the point; and y the distance from the horizontal axis, i.e., the vertical coordinate. It is shown in geometry that if (x_1, y_1) and (x_2, y_2) are vectors (or points in the plane), then the vector $a(x_1, y_1) + b(x_2, y_2)$ represents a point which lies on a straight line connecting the two points. In particular, if the sum is a weighted average, i.e., $a + b = 1$, then the point lies between the two points, nearer to the point with the greater weight.

Vectors with n components can be viewed as points in an n-dimensional space. Consider now a *set* of vectors $(x_1, x_2, \ldots, x_n)$ defined by the condition that $\sum_{i=1}^{n} x_i = c$. That is to say, the x_i are free to vary independently, subject only to the restriction that their sum shall be a fixed number c (a *constant*). If $n = 2$, the set of points represented by these vectors will all lie on a straight line. If $n = 3$, they will all lie on a plane. In general, the points of such a set will all lie in a space of one dimension less than the original space.

The most elementary geometrical figure lying in a space of a given number of dimensions is called a simplex. In this book, we shall mean by a 2-simplex, a triangle; by a 3-simplex, a tetrahedron, etc. We shall have occasion to examine simplexes defined by the following restrictions:

$$\sum_{i=1}^{n} x_i = 1;\ x_i \geqslant 0 \quad (i = 1, 2, \ldots, n). \qquad (0.51)$$

If $n = 3$, such a simplex, being a 2-simplex, is a triangle formed by the intersection of the plane

$$x + y + z = 1 \qquad (0.52)$$

with coordinate planes, $x = 0$, $y = 0$, $z = 0$. All the points on the sides of and inside the triangle constitute the simplex.

We shall have occasion to refer to simplexes in the

following contexts. A vector can represent a disbursement of payoffs to each player in an N-person game. At times we shall want to talk about sets of such vectors as sets of all possible disbursements that can obtain under certain conditions. Sometimes these sets will be simplexes, sometimes unions or intersections of simplexes. Note that simplexes are sets (of points).

Consider the set of points (x, y) in two-dimensional space defined by the condition that $y = f(x)$ where $f(\ )$ is a specified function of x. In particular, let $f(x) = 4x - x^2$. These are the points which *satisfy* the equation $y = 4x - x^2$; i.e., those whose coordinates, substituted for x and y respectively, turn the equation into an *identity*. Thus the coordinates of the points (0, 0), (1, 3) and (17, −221) satisfy the equation, but those of the point (1, 1) do not. It turns out that all of the points which satisfy the equation lie on a curve drawn in the plane (in this case a parabola). We may wish to find the particular point (x^*, y^*) which has the largest value of y. It turns out there is such a point, namely (2, 4). It is at the vertex of the parabola. The procedure for finding such maxima is described in differential calculus. It involves *differentiating* the function (finding its derivative), and setting it equal to zero, and solving the resulting equation for x. The resulting value or values may be maxima or minima (sometimes neither); but in a procedure involving further differentiations, it is possible to determine which are which. Thus the derivative of $4x - x^2$ turns out to be $4 - 2x$, which, when set equal to zero, yields $x = 2$. The second, derivative turns out to be -2, which, being negative, guarantees that $x = 2$ is a maximum.

How to Read the Mathematics in This Book

It remains to apply these ideas to interpreting the notation used in this book. As has been said, a principal difficulty in understanding a mathematical discussion is the unfamiliarity of the notation. I believe that in the case of game theory, it is *the* principal difficulty.

In Chapter 2 (p. 85), we have the equation

$$v''(S) = cv'(S) = c\left[v(S) + \sum_{i \in S} a_i\right]$$

$$= \frac{v(S) + \sum_{i \in S} a_i}{v(N) + \sum_{i=1}^{n} a_i}. \qquad (0.53)$$

Reading from left to right, we first encounter $v''(S)$, which (as appears from the text) is a value of a characteristic function of a certain game. The double prime should alert us to the likelihood that this function was obtained from another characteristic function. And so it was, as denoted by $cv'(S)$; namely by multiplying all the values $v'(S)$ by the same number c. Also the values $v'(S)$ were obtained from the original values $v(S)$ by adding to each value $v(S)$ a sum of numbers a_i, each being associated with the players in the coalition S, as indicated by $i \in S$ under the summation. Moreover, the number c is determined by adding to $v(N)$ (the value of the game to the whole set of players) a sum of numbers, each associated with a player of the game, and taking the reciprocal of the result.

The purpose of transforming the characteristic function in this way will become clear in the discussion of this matter.

Sometimes it is advisable to read a mathematical expression "from inside outward." Take the expression explained on p. 27:

$$\operatorname*{Max}_{i} \operatorname*{Min}_{j} (x_{ij}). \qquad (0.54)$$

Inside the parentheses we have a symbol which stands for any of m × n where i ranges over m numbers and j over n. Next, $\operatorname*{Min}_{j}$ tells us to take the smallest x_{ij} among those with a *fixed* i, as j varies over its range. Suppose it is a particular one, and call it x_i^*. There will be m of these, one such for every i. Finally, $\underset{i}{\text{Max}}$ tells us to take the largest of these as i varies over *its* range. The same principle can be applied to expression (11.8) on page 179, where the role of the indices i, j is played by the sets S and $-$ S.

To take another example, let us read expression (7.26) on p. 131:

$$\text{Max}\ [0, v(\overline{13}) - x_1]. \qquad (0.55)$$

Here $v(\overline{13})$ is a number determined by the coalition $(\overline{13})$ of players 1 and 3, x_1 is the payoff to player 1, and $v(\overline{13}) - x_1$ is their algebraic difference, which may be positive or negative. We have, therefore, within the brackets *two* numbers, namely 0 and this difference. "Max" tells us to take the larger of the two, namely 0, if the difference is negative; the difference, if it is positive.

In general, if there are parentheses within other parentheses (to distinguish—the latter are usually braces), within still other parentheses (usually brackets), it is a good idea to read from inside out.

A set is usually denoted by braces. Inside the braces may be an enumeration of the members of the set. (If these are themselves sets, they will be in other types of parentheses.) However, the braces may include a *definition* of a set. An example is expression (7.6) in Chapter 7:

$$\mathfrak{J}_{k,l} = \{D: D \subset N, k \in D, l \notin D\}. \qquad (0.56)$$

It defines *any* set D *such that* (denoted by :) D is included in a given set N, *and* k is a member of D, *and* l is not a member of D. The braces indicate that we are defining the *set* of all such D's. And since any such set is specified, once k and l are specified, k and l appear as subscripts of $\mathfrak{I}_{k,1}$. $\mathfrak{I}_{k,1}$ is thus a set of sets. In this book, sets of sets are usually denoted by script capitals.

It will be helpful to keep in mind that parentheses (of all kinds) serve two purposes: to include members of sets, and to indicate arguments of functions.

Turning to Chapter 5, let us interpret equation (5.11).

First, let us interpret the expression in the bracket. $v(S)$ is the value of the game to a certain (unspecified) coalition S. Now the *set* $S - \{i\}$ is a set of players from which the set containing only player i has been subtracted. Note that in subtraction, the sets must be of the same logical type; therefore player i must be viewed as a set containing that player only. $v(S - \{i\})$ is the value of the game to the coalition defined by the set $S - \{i\}$. The expression in the brackets is the difference between the two values, hence a number, which is specified once S and i are specified.

Next, $n!$ (read "n *factorial*" denotes the product $n(n-1)(n-2) \ldots (2)(1)$, and so for every integer, in particular $(s-1)$ and $(n-s)$. For example, $(n-s)! = (n-s)(n-s-1)(n-s-2) \ldots (1)$. The integers n and s denote the numbers of members of N and S respectively.

Hence the entire expression to the right of Σ denotes a number specified by S or i. Now we hold i fixed and sum all such numbers as S varies over all the subsets of N (including N). The result is the so-called Shapley value (cf. Chapter 5) of the game to player i, as denoted by the subscript on ϕ. If we now vary i, we find the Shapley value of the game to every player.

Turning to Chapter 9, let us spell out the meaning of equations (9.2)–(9.4). Here F stands for functions of a certain set. The particular function of the set, denoted by the subscript on F, is determined by how the players have partitioned themselves into coalitions. The partitions are displayed in the previous set of equations (9.1). The argument of each function F is a set of players. Thus

(i) denotes a single player, (ij) a pair. The value of each function is given on the right. Thus equation (9.3) says:

"If the players have partitioned themselves so that one player (i) is alone while the remaining two (j, k) are in a coalition, then the value of the game to i is 2, and to the coalition 0. This is true whoever is the single player."

Let the reader now test his "mathematical literacy" by interpreting equation (13.3) of Chapter 13. If its meaning is clear, he will encounter no further difficulty in understanding the mathematical notation in this book.

Technical points not covered in this Introduction will be referred to in Notes at the end of the volume.

PART I. *Basic Concepts*

1. *Levels of Game-theoretic Analysis*

The origins of game theory stem from concerns related to rational decisions in situations involving conflicts of interest. The term game theory itself derives from the analysis of so-called *games of strategy* such as Chess, Bridge, etc. Serious research in this area was doubtlessly stimulated by a need to bring to bear the power of rigorous analysis on problems faced by persons in the culturally dominant roles of "decision makers." That connections between games of strategy and strategic conflict already exist in the minds of men of affairs appears in the metaphors linking the languages of business, international relations, and war; in short, spheres of activity where strategic shrewdness is assumed to be an important component of competence.

Thus it is easy to portray game theory as an extension of a theory of rational decisions involving calculated risks to one involving calculations of strategies to be used against *rational* opponents, competitors, or enemies; that is, actors who are *also* performing strategic calculations with the aim of pursuing *their* goals and, typically, attempting to frustrate ours.

In short, the game metaphor (business is a game, life is a game, politics is a game) is already firmly established among people whose careers depend on the choice of right decisions and among those who have an appreciation of this process. Consequently, the above mentioned definition of game theory is easily related to what people already know, understand, and appreciate.

Unfortunately, inferences likely to be made from the definition easily lead to a misconception about the scope and the uses of game theory. The widespread appreciation of the decision maker's role makes it easy to put oneself into his shoes. If I were a decision maker (one might, perhaps, ask oneself), what would I expect from a theory which purports to deal with rational decisions in conflict situations? Clearly, I would expect from such a theory some indications of how rational decisions are to be singled out from all the available ones. Certain features of rational decisions follow from common sense considerations. One must know the range of the possible outcomes which can result as consequences of one's own choices and also of choices made by others, as well as, perhaps, certain chance events. Having made a list of these outcomes, one must know one's own order of preference among them and, as appears after a moment of reflection, also the orders of preference of *other* decision makers, who also exercise partial control over the outcomes by their own choices. Next, one must distinguish between immediate, intermediate, and final outcomes of decisions. Often it is difficult to decide what the preferences are (even one's own) with regard to the immediate and intermediate outcomes. For these can be evaluated only with reference to the final outcomes, to which they eventually lead, via additional choices of all concerned; and the relation between the former and latter is often not clear.

Anyone who has reflected on problems of decision making in conflict situations will usually quickly grasp the significance of these issues. Namely, these issues must be faced if game theory is to be a useful tool in the search for rational decisions. At this point, the game theorist who has undertaken to explain what he is about faces the difficult task of shifting attention *away* from these issues, which are not the central ones in game theory, toward other much more fundamental issues. He will have to explain the difference between the theory of some *specific* conflict situation and a *general* theory of such situations.

Let us first see what might be the shape of a theory of a specific, strictly formalized conflict situation, say a game of Chess. In Chess, the immediate outcomes of choices are "positions," i.e., the dispositions of the pieces on the board. These positions are controlled partially by one player, partially by the other, specifically, by the choices which the players make alternately. The positions must be evaluated only with reference to the bearing they have on the final outcome of the game; that is, on whether the outcome is a win for White, a draw, or a win for Black. The preference order of the players among *these* outcomes is clear. White prefers them in the order named; Black prefers them in the opposite order. Thus the *only* real problem the chess player faces is that of estimating the bearing which the immediate and the intermediate situations (the "positions") have on the final outcome.

Indeed, the theory of Chess deals entirely with this problem. In the theory of Chess, certain frequently occurring situations are analyzed with a view of deciding to which of the three possible final outcomes they are likely to lead. Sometimes analysis yields a definitive answer. It is known, for example, that if all the pieces have been captured except one rook, then the possessor of the rook *must* win the game provided he guides his choices by certain specified rules. Definitive prognoses can be made also in other more complex "end game" situations (i.e., when only a few pieces and pawns remain). Prognoses on the basis of situations arising in the beginning or the middle of the game are, as a rule, not definitive. But this is because the analysis of such situations is too complex to allow the investigation of all possible lines of play. Nevertheless, some prognoses can be made with considerable confidence on the basis of several centuries of experience. Other prognoses are noncommittal except to the extent of assertions like "White (or Black) has a better position." These too are based on intuitive judgments derived from experience of able players.

Chess theory is, then, essentially concerned with the search for "effective strategies," i.e., with the search for choices which are "likely" to lead to definitely winning positions or to prevent the opponent from achieving them. It is important to note that "likely" in this instance is not to be understood in the sense that Chance intervenes in the development of the situations the way it does in games of chance. Chance has "nothing to say" about what positions will emerge in a game of Chess, since each position results from a deliberate choice by a player.[5] The term "likely" is unavoidable in the formulation of Chess theory, because the situations are too complex to be analyzed in their entirety. It is conceivable that, if a complete analysis were carried out, the outcome of *every* Chess situation could be predicted with certainty (assuming that such analysis could be carried out by both players). In fact the end game situations mentioned above are precisely such; and so are Chess problems, whose solution depends on complete analysis. Such situations *can* be analyzed completely, and for this reason, when they arise in games played by experienced players, the game is broken off, the outcome being known to both. Therefore it makes sense to conceive of "progress" of Chess theory in terms of subjecting more and more situations to deeper and deeper analysis. It follows that the Chess player able to pursue such analysis more deeply acquires thereby a greater control over the situations and is in a better position to win the game.

All this makes good sense to a decision maker faced with choices in situations involving a conflict of interest. He may be well aware that the situations with which he is faced are far from being as clear cut as a game of Chess. The range of alternative choices (especially those available to an opponent or opponents) may not be known with certainty, nor the opponents' preference orders for the outcomes. It may be difficult to decide whether situations are to be assigned the status of intermediate or final outcomes and so to decide whether they are to be evaluated as "means to an end" or as "ends," etc. Nevertheless, a decision maker sophisticated in the ways of science (where concrete problems must always be translated into simplified or idealized "models" and where hypothetical assumptions must always be made because knowledge is never complete) can conceive of conflict situations which, under certain conditions, can be formulated as well-defined "games." If so, he may be willing to examine the sense in which game theory can be relevant to the problem of choosing rational decisions.

After all the preliminary conditions have been fulfilled, what *is* the problem from the decision maker's point of view? It is that of pursuing analysis sufficiently far so as to single out strategic choices which will either bring about the preferred outcomes or are "likely" to bring them about ("likely" in the estimation of persons with experience in similar situations).

In other words, it seems to the decision maker quite natural to see the task of game theory as a generalization of the task posed by the theory of Chess: the search for effective strategic decisions (after the problem has been sufficiently defined). Therein lies a misconception, because the search for effective decisions is *not* a central problem of game theory.

Game theory (as developed by people who have come to be recognized as game theorists) is properly a branch of mathematics. To the extent that many problems of mathematics are (or, at least, have been) instigated by abstractions from real life situations, game theory, too, can be so viewed. However, the mathematicians' research tools are different from those of the natural scientist (who deals with the world perceived by the senses). Consequently the problems usually posed by the mathematician are typically not at all the problems posed by "life," although they may have been instigated by impressions gathered from life.

The mathematician pursues his science by ascending to ever higher levels of generality, hence of abstraction. It seems natural to suppose that by solving a "more general class of problems" than those originally posed, the mathematician is thereby enabled to solve also the original problems; for does not the general case embody the special case? It does indeed happen that by ascending to a higher order of abstraction the mathematician is enabled to solve the problem originally posed and all the other problems of the same type. Frequently, however, this change of perspective has a different consequence, namely, the *abandonment* of a class of problems in favor of another class *which may never have arisen* in the original context.

As an elementary example, consider the problem of solving an equation with one unknown. At first, solutions of special cases of such equations were found; then a general method of solving such problems. Using this method, one could, of course, solve any special case.

However, in the process of establishing a method of solving linear equations in the first degree, certain problems appeared which had nothing to do with the problems that initiated this search. A special problem of solving a first degree equation might have arisen in the context of searching for a specific number among a set of known numbers. For example, a money changer, in converting one currency to another and charging a fixed fee for his service, might want to solve a linear equation. If he sells shekels at the rate of 350 minas per shekel and charges 4 minas for performing the transaction, he may want to know how many shekels he should give for 1000 minas. The equation to solve for x is

$$350x + 4 = 1000. \qquad (1.1)$$

The problem "has an answer" in terms of the operations which the money changer habitually performs, because problems "without an answer" do not as a rule arise in this context.

The mathematician, however, having posed the problem of solving the general linear equation will very quickly come across problems which have "no solutions" in the hitherto conventional sense. In order to make such problems solvable (so as to "round out" the theory), the mathematician invents new number entities. In the context of making all linear equations solvable these are negative numbers. Once this is done, new questions arise of purely mathematical nature; for instance, how are the usual operations of arithmetic to be extended so as to include negative as well as positive numbers? To be sure, contexts in which negative numbers acquire meaning were not long in appearing, e.g., credit accounting. Nevertheless it is quite possible that the concept of negative numbers arose in the purely mathematical context before this concept was put to work in "real life" applications.

The importance of the purely mathematical context becomes much more important in the theory of the quadratic equation. Here "answers" occur which cannot possibly arise in the context of either counting or physical measurement, namely, irrational and imaginary numbers. Hence a large part of the theory of quadratic equations is concerned with matters other than finding "desired magnitudes" as answers to problems posed by life. For instance, the question of *when* a quadratic equation has rational roots, real roots, two distinct roots, etc., can be asked quite apart from the problem of *finding* these roots.

As the restriction on the degree of an algebraic equation is removed, the theory becomes even more general. A part of the theory of equations is concerned with certain mathematical systems (called groups) in which the elements (that is, the entities operated upon) are no longer numbers but are themselves operations (called automorphisms). This branch of mathematics later turned out to have extensive applications, for example in crystallography and in quantum mechanics. However, in the process of its development, the original problems which have given impetus to the theory of equations (from which group theory arose) were completely lost sight of. That is, investigations in group theory have next to nothing to do with the solution of algebraic equations in the sense of finding magnitudes which satisfy it. For this reason, the term "theory of equations" may be misleading to someone whose attention is riveted on the original "practical" problems, in the context of which the beginnings of the theory were rooted. Nevertheless the general theory of equations sheds a brilliant light on the "nature" of algebraic equations and brings into focus certain of their aspects which are fundamental in many different branches of mathematics.

To give another even simpler example, the desk calculator was invented for the purpose of quickly adding long columns of whole numbers. It became the ancestor of the electronic computers. A branch of mathematics eventually was developed which deals with the design of such computers. A specialist in this science has neither a special competence nor the slightest interest in adding long columns of figures.

So it is with game theory. The preoccupations of game theorists have next to nothing to do with the problem of finding "effective strategies" in conflict situations. They have to do with matters which shed light on the "logic" of such situations. This logic turns out to be intricate and often perplexing, at times ridden with paradoxes, which, when resolved, provide us with insight concerning matters which had been either ignored or only vaguely understood.

Like all other branches of mathematics, game theory grew by progressive generalization, hence abstraction. Three levels of abstraction are clearly discernible: the theory of games in extensive form, the theory of games in normal form, and the theory of games in characteristic function form.

The fundamental "mathematical object" in the theory of games in extensive form is the so-called *game tree.* The game tree is determined by the rules of the game. If the game is to be represented by a game tree, the following must be specified in the statement of the rules.

1. A set of players.
2. A set of alternatives open to each player when it is his turn to make a choice among such alternatives (i.e., when it is his *move*). This set of alternatives will usually depend on the situation, which, in turn, is determined by the choices already made by all the players on their respective moves.
3. A specification of how much a player can know (when it is his move) about the choices already made by the players on previous moves.
4. A termination rule indicating situations which mean that the game is over.
5. A set of payoffs (one to each player) associated with every outcome of the game, the outcome being the situation in which the game has terminated.

The "root" of this tree can be represented by a point. The "branches" issuing from this point represent the alternatives open to the player who moves first. The end points of these "first order" branches represent the several situations which can result from the choices made by the first player. From these points, in turn, branches issue, which represent the alternatives open to the player who is to move next. (The identity of this player may also depend on the situation.) This branching process continues until a situation is reached which is defined (by the termination rule) as an outcome of the game.

If the player is always in a position to know the actual choices made by all the players who have already moved, then the game is called a *game of perfect information.* In such a game, each player is always in a position to know exactly to which branch point the play of the game has progressed. If such is not the case, the player can only know to which *set of branch points* the game has progressed, but not to which particular branch point of this set. These sets are called *information sets.* The rules of the game must be such that the alternatives open to a player at each point of an information set are in one-to-one correspondence; otherwise, not knowing to which point the game has progressed, he cannot know what alternatives are open to him; i.e., which choices are "legal."

A game tree is thus a diagram which specifies all the branch points, all the branches (labeled by the players who make the corresponding choices), and all the information sets. The game tree constitutes the representation of the game in extensive form. It is essentially a diagrammatic representation of the rules of the game. The theory of games in extensive form deals with the mathematical properties of game trees. A sample game tree is shown in Figure 2.

The most important concept emerging from the analysis of a game tree is that of a strategy. A strategy is essentially a statement made by a player specifying which of the alternatives he will choose if he finds himself in any of the information sets which are associated with his moves. It is shown in game theory that, once a strategy is chosen by each of the players, an outcome of the game is thereby determined.

One of the players in a game may be Chance. The alternatives open to this player are the possible states of a chance device, or one of several such devices, specified in particular situations. Thus the six faces of a die, the 38 positions of a roulette wheel, or the 52! arrangements

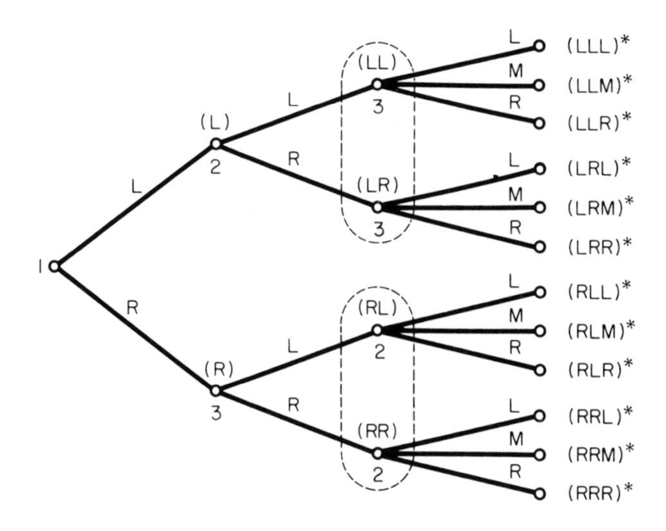

FIG. 2. A game tree representing a game with the following rules: Player 1 moves first, having a choice between "Left" (L) and "Right" (R). If player 1 chooses Left, player 2 moves next, having the same choice. If, however, player 1 chose Right, player 3 moves next, having the same choice. After two moves, the player who has not moved makes the last choice, and he has three choices, "Left," "Middle" (M), and "Right". This game is a game of perfect information. It would not be a game of perfect information if, for example, player 2's or player 3's choice on the second move were not known. In that case, player 3, if his were the last move, would only know that he is either at branch point (LL) or at (LR), but not where specifically. Thus his choice of Left might terminate the game either at (LLL)* or at (LRL)*. Player 2 would be in a similar situation. This is indicated by the dotted lines enclosing the information sets {(LL), (LR)} and {(RL), (RR)}. Note that player 1's choice must be known; otherwise players 2 and 3 would not know whose move it was following player 1's choice.

The numbers at the branch points indicate the player who is to choose; the branch points are designated by the choices already made (in parentheses); the final outcomes are in starred parentheses; the branches are designated by the choices which they represent.

To define the game completely, a triple of payoffs (one to each player) must be associated with each outcome. These can be assigned arbitrarily, and they are assumed to be known to all players.

of a deck of cards may be the alternatives among which Chance chooses.

If Chance is not a player, then, once a strategy is chosen by the players, the outcome of the game is determined. If Chance is a player, and the probabilities of her choices in the various situations are all known, then, once strategies have been chosen by all the bona fide players, the frequency distribution of all the possible outcomes of the game becomes known; hence also the *expected outcome* (as a statistical mean).

It is important to note that a strategy already contains as much "conditionality of choice" as the game allows. The degree of "conditionality of choice" is roughly the degree of dependence of a player's choices on the situation in which the choice is made. It is thus related to the degree of "flexibility" which characterizes a player's performance. Usually one associates such flexibility with rational decisions, that is, decisions which take into account the special circumstances in which they are made. In practice, a "flexible" player defers decisions until the relevant situation obtains. However, completely flexible decisions can also be made far in advance by specifying choices in all the foreseeable circumstances which may occur. A strategy, by definition, involves foreseeing *all* the possible situations which may arise in the course of a game. Their number is, to be sure, super-astronomical in all but very trival games. The "rational player," as he is defined in game theory, has unlimited memory capacity and unlimited skill of computation. Hence, by choosing a strategy before the game begins, he is already exhibiting as much flexibility as is possible under the rules of the game. For this reason, the "rational player" gains nothing in flexibility by deferring decisions.

If the object of game theory were to uncover effective strategies in situations involving conflicts of interest (the most easily understood "practical application" of game theory), then the investigations would have to be centered (1) on the construction of the game tree and (2) on examining the outcomes resulting from the combinations of strategies by the several players. Enormous difficulties would be encountered here. To see this, let us take one of the simplest games of strategy—Tic-Tac-Toe, and see what is involved in constructing its game tree.

Issuing from the root there are 9 branches, which represent the 9 alternatives open to player 1. Each of the next branch points will have 8 branches. We must continue this branching process to at least 5 moves, since no game can end before the fifth move. By the time we get to the fifth move, we have $9 \times 8 \times 7 \times 6 \times 5 = 15{,}120$ branches.

To be sure, we can reduce this number drastically by taking into account the symmetries of the Tic-Tac-Toe grid. For example, on his first move, player 1 has essentially only 3 alternatives: center, corner, and side. If he chooses center, player 2 has essentially two alternatives: corner or side; if player 1 chooses side, player 2 has essentially five alternatives (since one degree of symmetry remains); etc. Nevertheless, even taking symmetries into account, we would have a rather large tree; and although some effort would be saved in reducing the number of branches, more effort would be needed to examine the sets of situations which are equivalent by symmetry.

When one notes how simple it is to analyze Tic-Tac-Toe completely without the benefit of a complete game tree, one wonders whether game theory has anything to contribute to the problem of finding effective strategies by examining the extensive form of the game.

When we are dealing with real life situations, a preliminary problem must be solved before the game tree is constructed. One must ascertain the "rules of the game." Rules are not to be thought of as restraints imposed on the players by mutual agreement (as is the case with actual games). The essential function of rules is to delimit and to specify the available alternatives and the several situations which can result from the players' choices. These specifications may be consequences of the life situation itself. This is particularly true when the game does not proceed "in depth"; that is, where the outcomes of the choices are evaluated immediately. If the number of "moves" and the number of alternatives available at each move are both quite small, it may be possible to list each player's available strategies.

The representation of the game by its strategies alone is called representation in *normal form.* Such a representation is shown in Figure 3.

Once the game is represented in normal form, the rules of the game become irrelevant. The rules are important (from the standpoint of game theory) only to the extent that they determine the structure of the game tree and through it the available strategies and the outcomes associated with the combined strategy choices.

Once the strategies and the associated outcomes have been listed, the game tree also becomes superfluous. However, in most games worth playing, this conclusion is of no practical significance: the task of listing the strategies is an impossible one. For this reason theories of particular games deal not with total strategies but with the situations which may arise in the course of a play of the game. These situations could be examined with reference to each other if a game tree were constructed. But the construction of the game tree is likewise out of the question for most games of interest. Therefore, game theory, neither in its extensive nor in its normal form, has much of consequence to contribute to the problem of seeking out effective strategies.

Wherein, then, lies the contribution of game theory? The answer to this question can be understood only if one abandons any preconceived notions one may have had about the main objectives of game theory (supposing it to be concerned with the search for effective decisions in conflict situations).

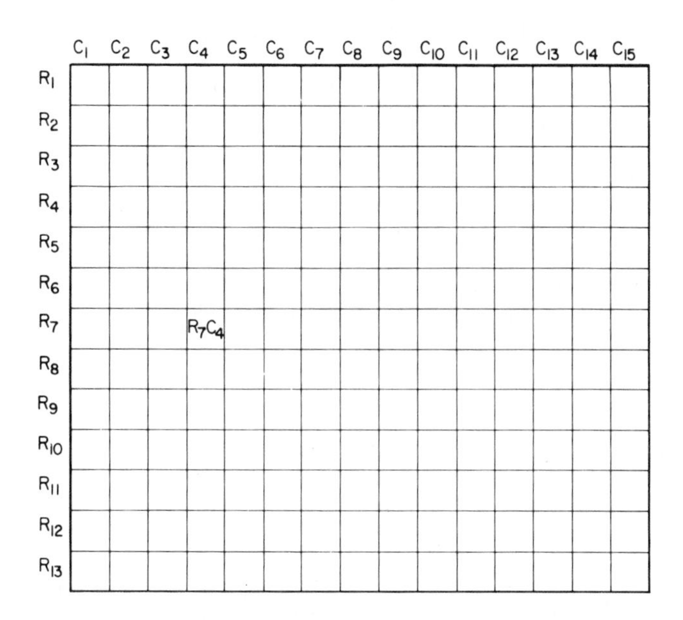

FIG. 3. A Two-person game represented in normal form. Here player 1 has 13 strategies available. They are represented by the horizontal rows of the *game matrix*. Player 2's 15 strategies are represented by the vertical columns. A play of the game consists of an independent (e.g., simultaneous) choice by each player of one of the strategies available to him. For instance, if player 1 chooses the strategy represented by R_7, and player 2 the strategy represented by C_4, the outcome of the game is the "box" of the matrix represented by (R_7, C_4). To represent the game completely, we must enter in each box a pair of payoffs, one to each player.

A Three-person game would have to be represented by a three-dimensional grid; an N-person game by an n-dimensional grid with an n-tuple of payoffs in each "box."

Once the game is represented in normal form, the game matrix rather than the game tree becomes the mathematical object of interest.

The normal form is most useful in the investigation of the Two-person game. In that case the matrix is a rectangular array, in which the strategies of one player are represented by the horizontal rows and those of the other by vertical columns.

From the standpoint of the theory of games in normal form, the simplest Two-person games are the so-called zerosum games, those in which the sum of the payoffs to the two players is always zero, regardless of the outcome. Practically all the so-called parlor games are represented as zerosum games, a reflection of the fact that what one player wins, the other loses (receives as a negative payoff). N-person games are also zerosum if the algebraic sum of the payoffs of all the players is zero regardless of the outcome, e.g., N-person Poker.

Of the Two-person zerosum games, the simplest (again from the standpoint of normal form representation) are those which have so-called *saddle points* in the game matrix. A saddle point is a "box" in the matrix in which the payoff to Row (the player whose strategies are represented by the horizontal rows) is minimal in its row and at the same time maximal in its column. Because the game is zerosum, it follows that the corresponding payoff to Column is the smallest in its column and the largest in its row. We shall say that a row (or a column) *contains* a saddle point if a box of the matrix in the corresponding row (or column) is a saddle point. Since a row (or column) represents a strategy, we can also say in that case that a strategy contains a saddle point. It is established in the theory of the Two-person zerosum game that if both Row and Column choose strategies containing a saddle point, the box determined by their choices will be a saddle point. (N.B. This is not necessarily true in games which are not zerosum, nor in N-person games.) Moreover, the payoffs in the saddle points of a Two-person zerosum game are all equal.

From this result it follows that the largest payoff which a player can get in a Two-person zerosum game with saddle points (assuming that each player is trying to get the largest payoff possible) is the payoff which accrues to him in a saddle point. Strategies which contain saddle points are called *maximin strategies.*

More generally, if x_{ij} is the payoff to Row in the i-th row, j-th column of the game matrix, then a strategy which contains $\underset{i}{\text{Max}}\ \underset{j}{\text{Min}}\ (x_{ij})$, regardless of whether it is a saddle point, is called a maximin strategy of Row; analogously a minimax strategy for Column.

Another important result concerns Two-person zero-sum games of perfect information (cf. p. 54). Every such game must contain at least one saddle point. Consequently, in every game of perfect information there exists at least one "best" strategy available to each player. If both choose such a strategy, the outcome of the game (in the sense of the associated payoffs) is determined in advance. If such "best strategies" are known to both players, it is pointless to play the game.

Chess, being a game of perfect information, must have saddle points in its strategy matrix. Therefore, if Chess were played rationally by both sides, the outcome of every play of the game would be the same: either White would always win, or Black, or every game would be a draw. The reason this does not happen is that Chess is too complex to be analyzed completely. It does happen in very simple games of perfect information like Tic-Tac-Toe or Nim.[6]

Note how the analysis of the game in normal form has led us away from the objectives of the analysis in extensive form. In the latter, the objective may well have been that of finding "effective strategies." (This objective is pursued in the theory of Chess, for example.) In the analysis of a game in normal form, the central question is whether a best strategy *exists* at all. Possibly the question would not even have arisen had not the analysis of games in normal form shown that in some games (e.g., those without saddle points which must be games without perfect information) one cannot define the "best" among the available strategies. In the case of games with saddle points, such best strategies do exist. However, the analysis contributes nothing to a practical method of finding such strategies in the original game (which is defined in terms of its rule).

The parallel with the theory of equations is instructive. A real life problem may induce us to seek a root of an algebraic equation. To have a physical meaning, such a root must be real. In the process of solving such problems, the theory of equations developed. This theory is more concerned with questions like whether an equation *has* a real root at all than with the problem of finding it. It turns out that sometimes an equation has, sometimes it does not have, a real root. (In particular, it always does if its degree is odd.) Clearly, if an equation has no real roots, it is futile to look for one. Similarly, if a game has no saddle points, it is futile to look for a "best" among the available strategies.

As has been pointed out, the development of mathematics has been marked by repeated extensions of the number concept, often instigated by the impossibility of solving certain equations within the framework of the older concept of numbers. An analogous extension of the concept of strategy was introduced into game theory to make all Two-person zerosum games (with or without saddle points) "solvable," if by a solution of a game we mean a pair of strategies which are somehow "best" for the respective players. This extension involves the notion of *mixed strategy*. A player chooses a mixed strategy if he uses some random device (e.g., a roulette wheel) to "choose" the strategy for him. Each mixed strategy is designated by a set of probabilities associated with each of the available strategies. To choose a mixed strategy means essentially to "fix" the device in such a way that strategy 1 will be chosen (by the device) with probability p_1, strategy 2 with probability p_2, etc. (Note that the choice of a *pure* strategy, i.e., a specific one among these available, amounts to setting its probability equal to one and all the other probabilities equal to zero.)[7]

The most important result in the theory of the Two-person zerosum game states that in every such game represented in normal form there exists a pair of strategies (pure if there is a saddle point and mixed if there is not) which are *in equilibrium* with each other. This means that if one player chooses such a strategy, then the other can do no better than to choose such a strategy also. Moreover, every such pair will be in equilibrium, and the expected payoffs will be the same in all equilibria.

The importance of this result, as the reader will probably concede, is not in its utility for seeking out "best" mixed strategies but rather in the further question it raises about the logical structure of games of various kinds. It turns out that all Two-person zerosum games are solvable in the sense that in them there always exist strategies in equilibrium with each other. Is this result valid for all Two-person games?

Once zerosum games have been defined, we can raise questions about games which are not zerosum; that is, about Two-person games in which the sum of the payoffs of the two players is not necessarily zero. It turns out that if this sum is always the same (even though not zero), the findings remain exactly the same. A little reflection shows that this cannot be otherwise. For, if the players receive jointly the same amount, regardless of the outcome, the game can be turned into a zerosum game by exacting this amount (if it is positive) or paying it (if it is negative) from (to) one of the players or from (to) both players in some proportion. This is equivalent to paying the players (or charging a fee) simply for playing the game (regardless of outcome), and so cannot make any difference in the strategic considerations governing the play of the game. Thus, to consider the games which are essentially different from zerosum games, we must turn to those where the sum of the payoffs to the two players is different in different outcomes of the game. These are the *non-constant-sum games*. Without loss of generality, they can be also called non-zerosum games.

In non-constant-sum games it may happen that some of the outcomes are preferred by both players to other outcomes (a situation which cannot happen in constant-sum games where the more one player wins, the more the other loses). In other words, in non-constant-sum games, the interests of the two players are in general not diametrically opposed.[8] They may be partially opposed and partially coincident. I do not know of any parlor games which are non-constant-sum, but, of course, real life situations are very common wherein the interests of parties partly coincide and partly conflict.

An important result in the theory of the non-constant-sum game in normal form states that every such game has at least one equilibrium (either in the sense of pure or mixed strategies). If one player "departs" from such an equilibrium while the other stays with it, the departing player cannot improve his payoff; he can only impair it if it changes at all. This result, however, is only partly analogous to the corresponding result in zerosum games, because it is not true that if both players choose a strategy (pure or mixed) containing an equilibrium, the outcome will necessarily be an equilibrium. Nor is it true that the payoffs to the respective players must be the same in all equilibrium outcomes. Accordingly, the theory of Two-person non-constant-sum games in normal form is more complex than that of constant-sum games.

Note the sense in which we use the word "complex." It has nothing to do with the complexity of the game in extensive form (e.g., the size of the game tree). Chess is enormously complex as a game actually played but, from the standpoint of the theory of games in normal form, Chess is almost trivial. It has a saddle point (being a game of perfect information); therefore each player has a best pure strategy, and this is all that game theory has to say about Chess. While the result says little that is of interest to Chess players, the attempt to answer similar questions in the context of non-constant-sum games reveals another "dimension of complexity," as it were. For example, it raises questions concerning what one can say about (non-constant-sum) games which are not "solvable" even though they possess equilibria. For, even assuming that rational players will choose strategies containing equilibria (even this assumption is open to question), we cannot be sure that the outcome will be an equilibrium.

Another question raised in the context of the non-constant-sum game in normal form is what the relation is between rational decisions and equilibria. In many non-constant-sum games both players can get more in some outcomes which are not equilibria than in others which are. In order to realize those outcomes which are better for both players, the two must choose their strategies *jointly*, and this requires some sort of agreement between the players. It stands to reason that if the players are rational, and if the situation allows the concluding of an agreement benefiting both players, it will be concluded. However, the coordinated strategies which benefit both players (in comparison with the uncoordinated ones) may be chosen in many ways, some choices favoring one player, some the other. The "solution" of the game then involves the determination of a reasonable compromise.

If there are more than two players, the situation becomes more complex. With two players, only one coalition (for the purpose of coordinating strategies) is possible, namely the coalition between the two players. Moreover, if the game is constant-sum, the coalition cannot benefit both players, since both cannot prefer one pair of strategies to another. If there are more than two players, even if the game is constant-sum, it may still be advantageous for some of the players to form coalitions. In that case, questions arise as to which coalitions can be expected to form if the players are rational, as well as the question about how the payoffs accruing to the coalitions will be apportioned among the members.

This leads to the third level of game theory, called the theory of games in *characteristic function form.* In this framework, the strategies available to the players are also abstracted from. The only givens of the game are now the payoffs which each of the several possible coalitions can assure for themselves respectively. Usually it is assumed that the payoffs are in some *conservative* and *transferable commodity* (such as money), so that it makes sense to add the payoffs and to speak of the *joint* payoffs accruing to each coalition. The specification of minimum joint payoffs which can accrue to each of the possible coalitions constitutes the characteristic function of the game.

It appears, then, that we have left the question of effective strategies far behind. Instead, other questions arose in the path of the developing theory, namely questions of *settlement* of the conflicts of interest among rational players. Note that in the simplest games (which from the standpoint of game theory are Two-person zerosum games) these questions of settlement do not arise. The interests of the players are diametrically opposed; the two players cannot gain anything by "reasoning together." Each knows that he can assure himself at least the value of the game to him, and knowing that the opponent can do likewise, knows what to expect. Thus the question how much each of the players will get is automatically settled. What then is left? Only the question of how to find the best strategies (pure or mixed). To answer this question, one must study the normal form of the game. To describe the strategies in terms of sequential choices conditional on situations, one must study the extensive form, for only the extensive form can actually display the specific decisions which constitute a strategy. This can be done only in the context of specific games.

In leaving these questions, we pose others on the higher levels of abstraction. In N-person games, for example, very complex questions arise about possible "settlements" of such games. This theory is sufficiently rich to present a challenge to the game theorist, quite aside from questions of identifying effective strategies. Mutatis mutandis, questions which are central in N-person game theory would have probably never been raised if the attention of game theorists were riveted on the search for "effective strategies."

If we take a cue from the history of mathematics, we see that very often it becomes necessary to abandon a class of problems, or rather to by-pass them, and to ascend to a higher level of abstraction where problems of an entirely different sort present themselves. From the vantage point of this higher level it often becomes apparent that some of the problems on the lower level were actually unsolvable in the framework of concepts characteristic for that level. Other problems, while solvable, become uninteresting, because from the vantage point of the higher conceptual level they can be reduced to routine operations. Game theory too proceeded along these lines, as it passed from the extensive form (where it did not tarry long) to the normal form, and finally to the characteristic function form representation of games. We shall begin our analysis with some very simple games in extensive form, then in normal form. Once we pass to the characteristic function form (and some of its modifications), we shall stay on that level of analysis; for there is where N-person game theory directs its main thrust.

2. *Three-level Analysis of Elementary Games*

We shall now examine some extremely simple N-person games close at hand, as it were, where their inner strategic structure is visible. Later we shall ascend to the higher rung, where the *structure* of strategies will no longer be seen. On that level, the strategies will become the indivisible units of analysis. Finally, we shall ascend to the highest rung, where the strategies are no longer visible and only the coalitions remain as the units of analysis.

The smallest number larger than two being three, and the smallest number of choices being two, the simplest N-person games in extensive form are those where each of three players in turn must choose between two alternatives. In other words, each player in sequence will have one move. The number of strategies available to the players will, however, depend on the information available after the moves are made. Thus, if the choices of the players become known after they are made, the player who moves second will have four strategies available, and the player who moves third will have sixteen. If the choices of no player are known to the others, then each player will have just two strategies.

Another important distinction among N-person games is in whether or not players are permitted to form coalitions. Finally we have the distinction between zerosum and non-zerosum games. We shall examine Three-person games in all of these variants. The first version will be a zerosum game of perfect information where coalitions are not allowed. This game is shown in extensive form in Figure 4.

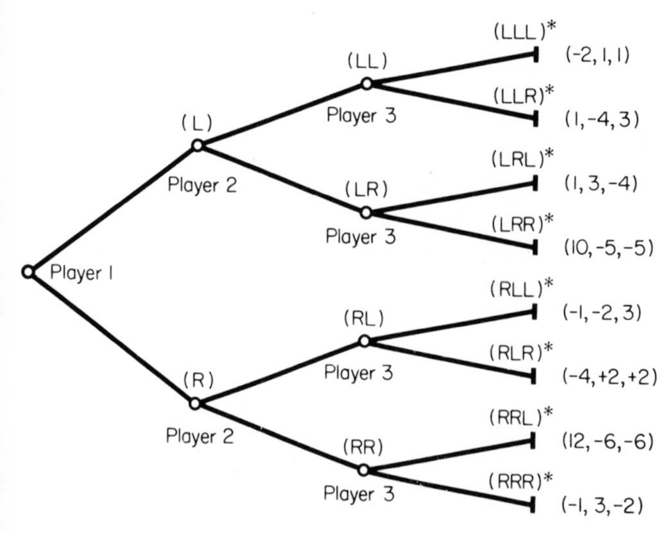

FIG. 4. "Left"-"Right" game as a game of perfect information. Each of three players moves in turn. Player 2 knows how player 1 has chosen; player 3 knows how both player 2 and player 3 have chosen. The branch points and the outcomes are designated by sequences of preceding choices. The payoffs are the triples in parentheses, accruing respectively to players 1, 2, and 3. Since the sum of the three payoffs is zero in every outcome, the game is zerosum.

Let us re-state the explicit meaning of "rational play."

1. Each player is able to foresee all possible consequences of each of his choices. Clearly, a particular ultimate consequence (i.e., the outcome and the associated payoffs) depends not only on how the player in question has chosen but also on how the others have chosen or will choose. For example, if player 1 chooses L, the outcome may be either (LLL) or (LLR) or (RLR) or (LRR), depending on how players 2 and 3 will choose. If player 2 chooses L after player 1 chose R, the outcome may be (RLL) or (RLR), depending on how player 3 will choose. An individually rational player is

supposed to scan all these possibilities in the process of making his decision.

2. Each player prefers the outcomes in the order of the magnitudes of the payoffs to him and to him alone. The payoffs of the other players are of no consequence to him *except* to the extent that knowledge about the others' preferences for the outcome helps him to anticipate the others' choices.

3. Each player assumes that every other player is individually rational like himself.

A strategy, we recall, is a plan of choices made by a player in which all possible contingencies, i.e., the choices to be made by the other players, are anticipated.

Let us first examine the game from player 1's point of view. He has just the two choices, L and R. Thereafter the game is out of his hands. In making his choice, he must consider how the others will choose in the pursuit of their interests. If player 1 chooses L, player 2 must choose between bringing the game to branch point (LL) or branch point (LR). In the former case, player 2 is sure to get -4 (since player 3 will choose R); in the latter case player 2 is sure to get 3, because player 3 will have to choose L, the lesser of two evils. Therefore player 2, being at branch point (L), will prefer R, and consequently the outcome of the game will be (LRL), which gives player 1 a payoff of 1.

Following the same reasoning for the other alternative, player 1 concludes that, in that case, the outcome will be (RRR), which gives him a payoff of -1. Consequently player 1 will choose L, and we have already seen that the choices of the other two players will realize the outcome (LRL). This will be the outcome if this game is played "every man for himself" under the condition of perfect information.

It is instructive to arrive at this result by examining the game in normal form. To do this, we must list the strategies available to each of the three players. Player

1 has just two strategies, [L] and [R] corresponding to his two choices, L and R. Player 2 has four strategies, since his two choices can be associated independently with each of the two choices of player 1. We shall introduce a positional notation to designate the strategies. Player 1's strategies will be denoted by the four pairs in square brackets:

$$[L, L]; [L, R]; [R, L]; [R, R]. \qquad (2.1)$$

The *position* of each member of the pair refers to player 1's choice; namely the first of the pair corresponds to player 1's choice of L, while the second member corresponds to player 1's choice of R. Thus player 2's strategy [R, L] means "If player 1 chooses L, choose R; if he chooses R, choose L." Strategy [R, R] means "Choose R regardless of player 1's choice"; etc.

Player 3 has sixteen strategies. They will be denoted by quadruples:

[L, L, L, L]; [L, L, L, R]; [L, L, R, L]; [L, R, L, L];
[R, L, L, L]; [L, L, R, R]; [L, R, L, R]; [R, L, L, R];
[L, R, R, L]; [R, L, R, L]; [R, R, L, L]; [R, R, R, L];
[R, R, L, R]; [R, L, R, R]; [L, R, R, R]; [R, R, R, R].

(2.2)

The entry in the first position is player 3's choice at the branch point (LL), the second at the branch (LR), the third at the branch point (RL), the fourth at the branch point (RR). Thus, strategy [R, R, R, L] reads: "Choose R except if the game is at branch point (RR), in which case choose L."

To represent a Three-person game in normal form we need a three-dimensional matrix, with the strategies of each player along each of the dimensions, in our case a $2 \times 4 \times 16$ matrix. A representation of a three-dimensional object on two-dimensional paper is awkward. To avoid it, we shall take advantage of the fact that player

	[L]			
	[L, L]	[L, R]	[R, L]	[R, R]
[L, L, L, L]	-2,1,1	-2,1,1	1,3,-4	1,3,-4
[L, L, L, R]	-2,1,1	-2,1,1	1,3,-4	1,3,-4
[L, L, R, L]	-2,1,1	-2,1,1	1,3,-4	1,3,-4
[L, R, L, L]	-2,1,1	-2,1,1	10,-5,-5	10,-5,-5
[R, L, L, L]	1,-4,3	1,-4,3	1,3,-4	1,3,-4
[L, L, R, R]	-2,1,1	-2,1,1	1,3,-4	1,3,-4
[L, R, L, R]	-2,1,1	-2,1,1	10,-5,-5	10,-5,-5
[R, L, L, R]	1,-4,3	1,-4,3	1,3,-4	1,3,-4
[L, R, R, L]	-2,1,1	-2,1,1	10,-5,-5	10,-5,-5
[R, L, R, L]	1,-4,3	1,-4,3	1,3,-4	1,3,-4
[R, R, L, L]	1,-4,3	1,-4,3	10,-5,-5	10,-5,-5
[R, R, R, L]	1,-4,3	1,-4,3	10,-5,-5	10,-5,-5
[R, R, L, R]	1,-4,3	1,-4,3	10,-5,-5	10,-5,-5
[R, L, R, R]	1,-4,3	1,-4,3	1,3,-4	1,3,-4
[L, R, R, R]	-2,1,1	-2,1,1	10,-5,-5	10,-5,-5
[R, R, R, R]	1,-4,3	1,-4,3	10,-5,-5	10,-5,-5

	[R]			
	[L, L]	[L, R]	[R, L]	[R, R]
[L, L, L, L]	-1,-2,3	12,-6,-6	-1,-2,3	12,-6,-6
[L, L, L, R]	-1,-2,3	-1,3,-2	-1,-2,3	-1,3,-2
[L, L, R, L]	-4,2,2	12,-6,-6	-4,2,2	12,-6,-6
[L, R, L, L]	-1,-2,3	12,-6,-6	-1,-2,3	12,-6,-6
[R, L, L, L]	-1,-2,3	12,-6,-6	-1,-2,3	12,-6,-6
[L, L, R, R]	-4,2,2	-1,3,-2	-4,2,2	-1,3,-2
[L, R, L, R]	-1,-2,3	-1,3,-2	-1,-2,3	-1,3,-2
[R, L, L, R]	-1,-2,3	-1,3,-2	-1,-2,3	-1,3,-2
[L, R, R, L]	-4,2,2	12,-6,-6	-4,2,2	12,-6,-6
[R, L, R, L]	-4,2,2	12,-6,-6	-4,2,2	12,-6,-6
[R, R, L, L]	-1,-2,3	12,-6,-6	-1,-2,3	12,-6,-6
[R, R, R, L]	-4,2,2	12,-6,-6	-4,2,2	12,-6,-6
[R, R, L, R]	-1,-2,3	-1,3,-2	-1,-2,3	-1,3,-2
[R, L, R, R]	-4,2,2	-1,3,-2	-4,2,2	-1,3,-2
[L, R, R, R]	-4,2,2	-1,3,-2	-4,2,2	-1,3,-2
[R, R, R, R]	-4,2,2	-1,3,-2	-4,2,2	-1,3,-2

FIG. 5. The left hand matrix corresponds to player 1's choice of strategy [L]; the right hand matrix to his choice of strategy [R]. The columns of each matrix represent player 2's four strategies; the rows player 3's 16 strategies. For example, let player 1 choose [R], player 2 [R, L], player 3 [L, R, R, L]. Then, since player 2's response to [R] is L and player 3's response to [RL] is R, the outcome is (RLR)* with the corresponding payoff vector (—4, 2, 2).

1 has only two strategies, and represent the game by two two-dimensional matrices side by side instead of one on top of the other. The first matrix represents the strategies of players 2 and 3 in the case that player 1 chooses his strategy [L]; the second in the case player 1 chooses his strategy [R]. The representation is shown in Figure 5.

Let us find the saddle point of this three-dimensional matrix. First we must eliminate the *dominated* strategies, i.e., those over which some player prefers another strategy as a "sure thing." Note, for example, that player 3's strategy [R, L, L, R] *dominates* strategy [R, R, L, L]. This is because player 3 gets as much or more in the former than in the latter regardless of what players 1 and 2 do. This can be verified by comparing player 3's payoffs (the third numbers in the triples) in the corresponding rows in each of the matrices shown in Figure 5. Being a dominated strategy, [R, R, L, L] should be eliminated from consideration.

In a similar way, we eliminate player 3's strategies [L, L, R, R], [L, R, L, R], [R, R, R, L], [R, R, L, R], [L, R, R, R] and [R, R, R, R]. Figure 6 shows the Left-Right game in normal form after the dominated strategies have been eliminated.

Next we examine all the remaining entries to see whether they are equilibria. They are not if any of the players can get a larger payoff by making an appropriate switch. For example, the entry in the upper left corner of the left-hand matrix is not an equilibrium, because player 1 can get −1 instead of −2 by switching to [R]. The intersection of [R], [L, L], and [L, L, L, R] is not an equilibrium, because player 2 can get 3 instead of −2 by switching to [L, R]. Having thus eliminated all the non-equilibrium entries, we see that the only ones remaining (the equilibrium entries) are those where the payoff vector is (1, 3, −4), which, we

[L]	[L, L]	[L, R]	[R, L]	[R, R]
[L, L, L, L]	-2,1,1	-2,1,1	1,3,-4	1,3,-4
[L, L, L, R]	-2,1,1	-2,1,1	1,3,-4	1,3,-4
[L, L, R, L]	-2,1,1	-2,1,1	1,3,-4	1,3,-4
[L, R, L, L]	-2,1,1	-2,1,1	10,-5,-5	10,-5,-5
[R, L, L, L]	1,-4,3	1,-4,3	1,3,-4	1,3,-4
[R, L, L, R]	1,-4,3	1,-4,3	1,3,-4	1,3,-4
[L, R, R, L]	-2,1,1	-2,1,1	10,-5,-5	10,-5,-5
[R, L, R, L]	1,-4,3	1,-4,3	1,3,-4	1,3,-4
[R, L, R, R]	1,-4,3	1,-4,3	1,3,-4	1,3,-4

[R]	[L, L]	[L, R]	[R, L]	[R, R]
[L, L, L, L]	-1,-2,3	12,-6,-6	-1,-2,3	12,-6,-6
[L, L, L, R]	-1,-2,3	-1,3,-2	-1,-2,3	-1,3,-2
[L, L, R, L]	-4,2,2	12,-6,-6	-4,2,2	12,-6,-6
[L, R, L, L]	-1,-2,3	12,-6,-6	-1,-2,3	12,-6,-6
[R, L, L, L]	-1,-2,3	12,-6,-6	-1,-2,3	12,-6,-6
[R, L, L, R]	-1,-2,3	-1,3,-2	-1,-2,3	-1,3,-2
[L, R, R, L]	-4,2,2	12,-6,-6	-4,2,2	12,-6,-6
[R, L, R, L]	-4,2,2	12,-6,-6	-4,2,2	12,-6,-6
[R, L, R, R]	-4,2,2	-1,3,-2	-4,2,2	-1,3,-2

FIG. 6. Strategies dominated by other strategies have been deleted.

have already seen, is the pre-determined outcome of this game.

So far the situation is quite similar to that in the Two-person game with a saddle point. If there is a saddle point, the players (after having eliminated the dominated strategies) can do no better than choose strategies containing the saddle point. Moreover, if each does this, the saddle point outcome will actually obtain.

N-person game theory differs in an important respect from Two-person theory in that the existence of strategies containing a saddle point does not guarantee that if such strategies are chosen by all players, the outcome will be a saddle point, that is, an equilibrium outcome. To see this, let us change the rules of our game so that now the players make their choices L or R in ignorance of how others have chosen or, what is the same thing, must make their choices simultaneously. In this version of the game, the number of strategies available to players 2 and 3 will be reduced. This is because, in the absence of information about how previous players have chosen, players 2 and 3 do not have "conditional" strategies at their disposal (strategies expressed as "If he chooses so and so, choose so and so"). Allowing only strategies [L] and [R] to each player, we have the normal form of this game as shown in Figure 7.

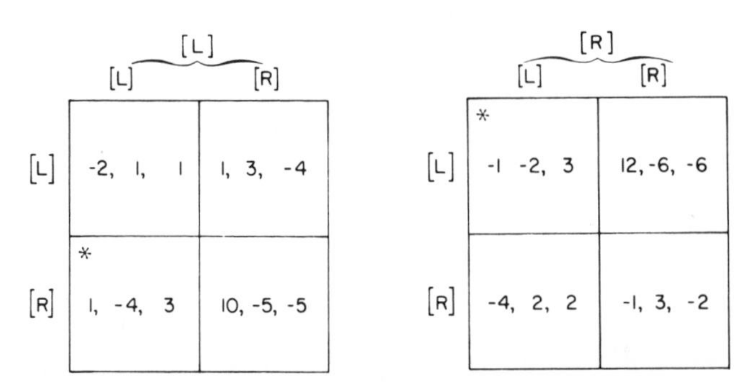

FIG. 7. Each player chooses in ignorance of the others' choices. Consequently each player has only two strategies. Equilibrium outcomes are marked by asterisks.

The saddle points are the boxes marked by asterisks. To choose a strategy containing a saddle point, player 2 must choose [L]. But players 1 and 3 may choose either strategy, because both contain saddle points. Suppose, then, all three players choose [L]. Player 1's choice puts the outcome into the left hand matrix of Figure 7; player 2's into the left column; player 3's into the upper row. The outcome is the left upper box of the left hand matrix. This is *not* a saddle point, hence not an equilibrium. It turns out that there is no way to prescribe a strategy *policy* (e.g., choose a strategy containing an equilibrium) to each player in a way that, if they make these choices, the outcome is certain to be an equilibrium. The concept of "individual rationality" becomes ambivalent.

To restore the concept of rationality in the context at least of the constant-sum game, we can allow the players to form coalitions. That is, we can let them coordinate strategies with each other so as to maximize their *joint* payoff.[9]

If a subset of the n players form such a coalition, and if the remaining players form a counter-coalition, we shall have essentially a Two-person constant-sum game, where the concept of rationality (now meaning the rationality of the coalition acting as a single player) is restored.

In this connection two comments are in order. First, it makes sense for the players to act in concert with the view of maximizing their *joint* payoff only if they can pool their payoffs as well as their strategies. We shall assume that this is the case. A situation of this sort may obtain if the payoffs are in some transferable, conservative commodity like money. Quantities of money can be added and apportioned among the members of a coalition. Second, we can reduce an N-person constant-sum game to a Two-person constant-sum game in more than one way, depending on which coalitions are formed.

Let us return to our game in its perfect information version to see what will be the outcome if different players join in coalition.

Suppose first that players 2 and 3 decide to form a coalition and so to act in concert in order to obtain the largest joint payoff they can.

Referring to Figure 4, we observe the following. If player 1 chooses L, players 2 and 3 can, by coordinating their choices, bring the game to the outcome (LLL), which gives them a joint payoff of 2, and consequently a payoff of -2 to player 1. If, on the other hand, player 1 chooses R, players 2 and 3 can bring the game to the outcome (RLR), which gives them a joint payoff of 4, and −4 to player 1. Player 1 has taken all this into account and will, of course, choose L. He cannot help losing 2 units, but he need not lose more. Consequently, if coalition $(\overline{23})$ forms against $(\overline{1})$, the *value* of the game to the coalition of two is 2, and to the coalition of one −2.

Suppose now that players 1 and 3 form a coalition against 2. In coordinating their strategies, they must take into account what 2 will do in response to each of player 1's choices. If player 1 chooses L, player 2 will choose L, because if he chooses R, he stands to lose 5 (since player 3 will choose R), but only 4 if he chooses L. Consequently, if player 1 chooses L, the coalition $(\overline{13})$ will get 4. On the other hand, if player 1 chooses R, thus forcing player 2 to choose L (to avoid a loss of 6), the coalition will get only 2. Therefore the coalition should begin the game by a choice of L by player 1. In this way the coalition will certainly get 4. The value of the game is 4 to the coalition $(\overline{13})$ and −4 to player 2 playing against this coalition.

Finally, suppose player 1 and 2 form a coalition. Coordinating their choices, they can bring the game to any of the branch points (LL), (LR), (RL), or (RR). Clearly, it is to their best advantage to bring the game to the branch point (LR), for then they will get at least 4, while player 3 gets −4.

In summary, player 1 playing against the coalition of the two others can get no more than −2 (but he need not lose more); player 2 playing against a coalition can expect to get −4, and so can player 3. Players 1 and 2 in a coalition can expect to get 4, and so can players 1 and 3. Players 2 and 3 can expect to get 2. As for the coalition of all three players, they are sure to get a joint payoff of zero, since the sum of the payoffs is zero in this game regardless of outcome.

A rule which assigns a value of the game to each subset of the n players is called the *characteristic function* of the game. By "rule" we mean here not a rule of the game but a consequence of the rules of the game in question; in other words, a rule in the sense of a mathematical function (cf. p. 30). The value of the game to each subset is defined as the value of a Two-person game which results if the subset in question plays as a single player and if the remaining players form a counter-coalition. If the N-person game is constant-sum, the value to each subset is defined unambivalently. If it is not constant-sum, there may be an ambivalence in the definition, as we shall see below. For the time being, we are discussing the characteristic function of constant-sum games only. We have seen that, without loss of generality, constant-sum games can be considered to be zerosum games. We shall so consider them.

By the way the characteristic function of a zerosum game is defined, we see that it must have the following properties.

1. The value of the game to the *grand coalition* (i.e., the coalition of all the players) must be zero, since the sum of the payoffs is zero regardless of outcome.

2. The value of the game to the null subset (cf. p. 17) must also be zero, since the players are not obligated to "throw away" any portion of their payoffs.

3. If S is a subset and −S the complementary subset (cf. p. 16), then the value of the game to −S is numerically equal to the value of the game to S but with opposite sign.

4. The value of the game to a union of two subsets (cf. p. 18) must be at least as large as the sum of the values to the subsets separately. This is because players joining in a coalition, hence coordinating their

strategies, can always do at least as well as if they had not joined in the coalition (assuming that the remaining players have joined in a counter-coalition and are attempting to get the largest joint payoff for themselves). This property of the characteristic function is called the *super-additive* property.

We express these properties of the characteristic function (of an N-person constant-sum game) mathematically as follows.

The characteristic function of an N-person constant-sum game is a function v( ) (cf. p. 33) on the set of subsets of N to the real numbers such that

$$v(\emptyset) = 0; v(N) = K, \text{ a constant} \tag{2.3}$$

$$v(S \cup T) \geqslant v(S) + v(T) \text{ if } S \cap T = \emptyset \tag{2.4}$$

$$v(S) = K - v(-S) \text{ where } -S \text{ is the complement of } S \text{ with respect to } N. \tag{2.5}$$

The game shown in Figure 4 (with coalitions allowed) has the following characteristic function:

$$v(\emptyset) = v(\overline{123}) = 0 \tag{2.6}$$

$$v(\overline{1}) = -2; v(-\overline{2}) = -4; v(\overline{3}) = -4 \tag{2.7}$$

$$v(\overline{23}) = 2; v(\overline{13}) = 4; v(\overline{12}) = 4. \tag{2.8}$$

Note that all the properties of the characteristic function are satisfied.

Let us now examine a Three-person non-constant-sum game to note some important aspects in which it differs from constant-sum games. A non-constant-sum game is shown in Figure 8.

The game will be recognized as a three-sided Prisoners' Dilemma game. The choices of the players have been labeled suggestively C (for cooperation) and D (for defection). If all three cooperate, each wins 1 unit. The Prisoner's Dilemma feature enters via rewards accruing to defectors, provided not all three

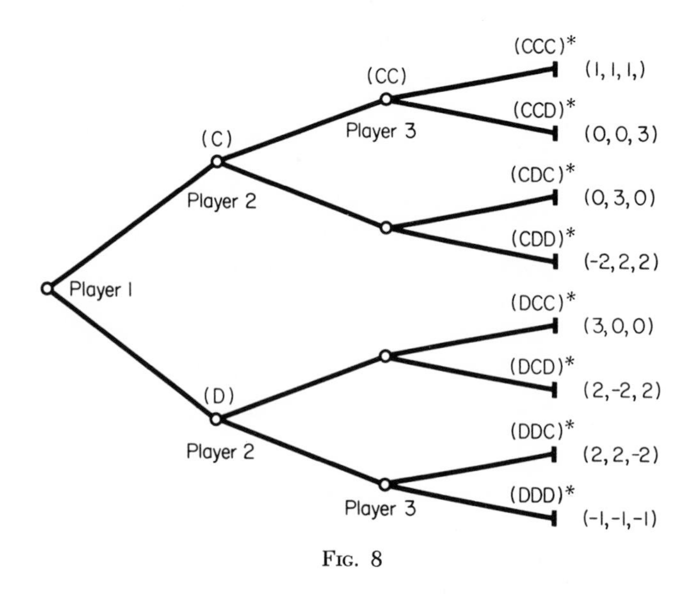

FIG. 8

players defect. In particular, a single defector gets the largest payoff 3. Each of two defectors gets 2, i.e., a smaller payoff than that of a single defector but more than that of a single cooperator or of each of two cooperators, 0. A single cooperator (the "sucker") suffers the largest loss, −2. If all three defect, each loses 1 unit.

If we analyze this game as before, assuming it to be a game of perfect information, we arrive at the single saddle point in (DDD), which is analogous to the saddle point in the Two-person Prisoner's Dilemma game. If, however, we allow coalitions in the game of perfect information, we get some curious results.

Consider first a coalition of players 2 and 3 versus player 1.

If player 1 chooses C, players 2 and 3 can realize outcome (CDD) to get a joint payoff of 4, whereby player 1 gets −2. If player 1 chooses D, the most players 2 and 3 can get jointly is zero, whereby player 1 can get at least 2. Therefore player 1 can get at least 2 playing against the (rational) coalition $(\overline{23})$.

If players 1 and 3 are in a coalition, player 2 can get at least 2. If player 1 chooses C, then player 2 will choose D, which will assure him at least 2. If player 1 chooses D, then player 2 will choose D. Since players 1 and 3 are in coalition, they will not make D player 3's final choice (for in that case they get jointly -2). Rather, player 3 will choose C, in which case the coalition will get 0 and player 2 will get 2.

If players 1 and 2 are in coalition, they can expect to get zero; for they can bring the game to any of the branch points (CC), (CD), (DC), or (DD), and in each case player 3 will make the choice which maximizes his own payoff. This will give zero to $(\overline{12})$ in all cases.

As for the grand coalition, it can assure itself a joint payoff of 3.

If we define the value of the game to each coalition as the amount which the coalition can expect to get playing against the rational counter-coalition, then the characteristic function of this game will be given as follows:

$$v(\emptyset) = 0 \tag{2.9}$$

$$v(\overline{1}) = 2; v(\overline{2}) = 2; v(\overline{3}) = 2 \tag{2.10}$$

$$v(\overline{23}) = 0; v(\overline{13}) = 0; v(\overline{12}) = 0 \tag{2.11}$$

$$v(\overline{123}) = 3. \tag{2.12}$$

We see that this characteristic function does not satisfy property (2.4) the super-additive property (cf. p. 79). At first thought, this may seem paradoxical. Why should not two players coordinating their strategies not do at least as well jointly as they can do separately? This is because in a non-zerosum game it may happen that, as two players cooperate to get a maximum joint payoff for themselves, they may (inadvertently) increase the payoff to the other player so that he gets even more than a coalition would get of which he might be a member.

In our example, player 1, knowing that players 2 and 3 (being rational) will be trying to maximize their joint payoff, not minimize his, can play D in the expectation that players 2 and 3 will not "punish him" by realizing the outcome (DDD), for they would be punishing themselves in the process. They will take 0 (the most they can get), and as a result, player 1 will get 2. We see that each of the three players is in a position to do the same, and so the single player has really an advantage in this game over any coalition of two players. It follows that no player will want to join in a coalition with just one other. Nor would he want to join with both others. The grand coalition can guarantee itself only 3 units, which, because of the symmetry of the game, would be expected to award only 1 unit to each. In short, every player wants to be in a coalition only with himself (assuming that the other two join in a coalition). But if each player remains alone, the outcome of the game, as we have seen, is (−1, −1, −1), and the purpose of not joining a coalition is defeated.

We see that the dilemma is even more severe in the Three-person version of Prisoner's Dilemma than in the Two-person version. In the Two-person version the dilemma disappears if the two join in a coalition, which is clearly in their interest to do. In the Three-person version it is not in the interest of either player to join in a coalition with one other (since a single player gets more than a coalition of two) nor with both of the others, since in that case he can reasonably expect only 1. Moreover, if a player remains alone, it *is* in the interest of the other two to join in a coalition, since in that case they can jointly get 0, while if every man plays for himself, each gets −1. Therefore each man will strive to remain out of a coalition, expecting that the other two will join in a counter-coalition. But we have seen that if each player remains alone, each player loses, whereas in a coalition of two he would break even, and in a coalition of three he could expect to get 1 unit as an equal share of the joint gain.

The paradoxical features of a non-super-additive characteristic function can be avoided by re-defining it for non-constant-sum games. Imagine that a fictitious $(n + 1)$st player has been added to the game. This player does not participate in the game in the sense of choosing strategies, but he does get payoffs. Namely, in every outcome of the game he gets a payoff which is numerically equal and opposite in sign to the sum of the payoffs of all the bona fide players. The resulting game is then a zero-sum game, and its characteristic function is super-additive. This way of defining the characteristic function is equivalent to assuming that every coalition can expect that the counter-coalition will "do its worst," i.e., will act in such a way as to minimize the joint payoff of the first coalition. Thus the value of the game to each coalition is supposed to be the very *least* it can get if it acts in concert (regardless of what common interest dictates to the counter-coalition). In our subsequent discussion we shall use the one or the other versions of the characteristic function in various circumstances.

We shall now assume that the characteristic function of a game is super-additive, that is, conditions (2.3) and (2.4) are satisfied.

N-person games can now be distinguished according to those where condition (2.4) is satisfied only by equalities, and those where some of the inequalities are *strict* (cf. p. 29). In the former case, there is no inducement for any of the players to join in a coalition with any of the others, since there is no joint gain in doing so. Such games are called *inessential,* and, from the point of view of the theory of N-person games in characteristic function form, offer no interest. The remaining games are called *essential.* In this book, we shall be dealing almost exclusively with essential games and will use the term N-person game to mean an essential game unless otherwise specified.

In an essential game, since at least one inequality in condition (2.4) must be strict, it follows that

$$v(N) > v(\bar{1}) + v(\bar{2}) + \cdots + v(\bar{n}). \qquad (2.13)$$

In other words, all of the players joining in a grand coalition can get jointly more than the sum of what they can get playing every man for himself against a coalition of all the others. We shall now show that we can, without loss of generality, denote each $v(\bar{i})$ $(i = 1, 2, \ldots, n)$ by 0 and $v(N)$ by 1.

Suppose first that each player pays a fee or is given a bonus *regardless of the outcome of the game* so as to make $v(\bar{i})$ always equal to zero; that is, an amount numerically equal and of opposite sign is added to each $v(\bar{i})$. Recall that the $v(\bar{i})$ are determined by the game itself, assuming that each player plays rationally against the others combined in a coalition which is trying to minimize his payoff. Therefore the player should play the same way whether or not he receives the bonus (or pays a fee), so that these side payments (or charges) from the outside do not change the strategic structure of the game. We now have the "same" game with $v(\bar{i}) = 0$ $(i = 1, 2, \ldots, n)$. Clearly $v(N)$ is also changed in the process; namely, $v(N)$ becomes

$$v'(N) = v(N) + \sum_{i=1}^{n} a_i, \qquad (2.14)$$

where a_i is the side payment (positive or negative) to player i.

Now let us change the units of the payoffs so that

$$v''(N) = cv'(N) = 1, \qquad (2.15)$$

where c is an appropriate constant.

The constant, c, which effects this change (amounting to no more than a change in the units of payoff) is

$$c = \frac{1}{v'(N)} = \frac{1}{v(N) + \sum_{i=1}^{n} a_i}. \qquad (2.16)$$

Moreover, the value of the game accruing to each subset of S now becomes

$$v''(S) = cv'(S) = c\left[v(S) + \sum_{i \in S} a_i\right] = \frac{v(S) + \sum_{i \in S} a_i}{v(N) + \sum_{i=1}^{n} a_i}. \tag{2.17}$$

Since none of the $v(\bar{i})$ is now negative, it is now true that

$$\text{if } S \supseteq T, \text{ then } v(S) \geqslant v(T). \tag{2.18}$$

In particular,

$$v(N) \geqslant v(S) \text{ for all } S \subseteq N. \tag{2.19}$$

Hence, in the re-formulated version the characteristic function of *every* essential game can be written:

$$v(\bar{i}) = 0 \quad (i = 1, 2, \ldots, n) \tag{2.20}$$

$$v(N) = 1 \tag{2.21}$$

$$0 \leqslant v(S) \leqslant 1 \text{ for all } S \subseteq N. \tag{2.22}$$

This so-called *normalized* form of the characteristic function makes it convenient to compare different games, and simplifies the theory.

Example

Examining the characteristic function of the game shown in Figure 4 we see that it is an essential game. In normalized form, its characteristic function becomes

$$v(\bar{i}) = 0 \quad (i = 1, 2, 3) \tag{2.23}$$

$$v(\overline{123}) = 1 \tag{2.24}$$

$$v(\overline{23}) = 1; v(\overline{13}) = 1; v(\overline{12}) = 1. \tag{2.25}$$

Let us see what happens when we normalize *any* essential Three-person constant-sum game. If $v(\bar{i}) = 0$ and $v(\overline{123}) = 1$, it follows by property (2.5) (cf. p. 79) that $v(\overline{12}) = v(\overline{13}) = v(\overline{23}) = 1$, since the payoff to any coalition of 2 plus the payoff to the third player must always add to 1. Therefore the theory of the essential constant-sum Three-person game in characteristic function form "collapses" into the theory of a single game. The game can be stated as follows. The three players are to divide a unit. Any two of them by forming a coalition can get the whole unit. This game has been used repeatedly as an example of a Three-person constant-sum game. We see that if we are interested only in the features of the game revealed by its characteristic function, then *every* essential Three-person game is equivalent to this example.

If there are more than two players, however, there is "more room" for variety. Only the value of the game to coalitions of n-1 players must be equal to 1 in the normalized form of the game. The value of the game to the several coalitions with more than one but fewer than n-1 players can be any numbers between 0 and 1.

If the game is not constant-sum, even more variety is possible. For then the restriction $v(S) = 1$ for all S with n-1 players need not be imposed.

An alternative way of normalizing an N-person game is by transforming the payoffs in such a way that $v(\bar{i}) = -1$ $(i = 1, 2, \ldots, n)$, $v(N) = 0$. Note that this notation does not imply that the game is necessarily zerosum; it is no more than a shift of "base line" and of the unit of utility. We shall sometimes use this alternative form of normalization.

In what follows we shall consider N-person games almost exclusively with regard to their characteristic functions (or generalizations thereof). Roughly speaking, the givens of the game will be what each subset of the n players can expect to get. The problems will concern mostly what each individual player can expect to get, and to some extent how the players can be expected to form coalitions. As we shall see, the answers to the first question are far from definitive, and to the second even less so. Much, however, can be learned in the process of posing and *attempting* to answer these questions, and therein lies the principal intellectual value of N-person game theory.

3. Individual and Group Rationality

A fundamental problem posed by game theory is to determine the outcome (or outcomes) of games which can be expected to occur if games are played by "rational players." Clearly, the problem acquires meaning only after the concept "rational player" is well defined.

When a mathematical concept is defined, it is by implication well defined, since mathematical concepts arise only in the process of logically precise definitions. However, when a concept arises outside the content of exact definitions, for instance, on the basis of our intuitive "understanding" of something, it turns out not infrequently that attempts to apply the concept lead to paradoxes or contradictions.

So it is with the concept of a "rational player." Intuitively, a rational player can be defined as one who, in making strategic choices, takes into account the consequences of his choices and of those made by other players, and is guided by the attempt to maximize the expected utility of the outcome (to him and to him alone) under the constraints of the situation. However, as is apparent from the analysis of non-constant-sum games, and as will become apparent from our analysis of N-person games, this definition is far from sufficient to decide whether a given strategic choice is rational. It seems that the concept of rationality bifurcates into "individual rationality" and "collective rationality," and the prescriptions derived from the one often fail to coincide with those derived from the other.

We have seen how this problem arises in Two-person games, for example, in Prisoner's Dilemma. It becomes even more complex in N-person games.

Given the characteristic function of a particular game, and assuming the players to be "rational," it is natural to inquire (1) which of the possible coalitions can be expected to form; and (2) what will be the final disbursements of payoffs among the players.

Recall that the payoffs have been assumed to be in some transferable and conservative commodity, so that side payments can be freely made among the players. It follows that the players can bargain about how they will divide a joint payoff accruing to a coalition among the members of the coalition. Thus, players can possibly be induced, by promises of such side payments, to join or to switch coalitions. Leaving aside for the moment the question of which coalitions are likely to form, we shall concentrate on the question of disbursing the payoffs.

Now we can define "individual rationality" and "group rationality" in terms of what individuals and groups of individuals organized into coalitions can expect to get.

First, we can safely assume that a rational player cannot be induced to accept any payoff which is less than the value of the game to him. This is because, in order to get the value, he need not bargain at all. He can simply refuse to join any coalition and play the game alone vis-à-vis the others. By definition of the value of the game, he can expect to get at least so much, even if all the other players form a coalition "against him" as in the constant-sum game. Let now any disbursement of payoffs among the n players be designated by the n-tuple numbers, the *payoff vector* (cf. p. 34):

$$\vec{x} \equiv (x_1, x_2, x_3, \ldots, x_n). \qquad (3.1)$$

The principle of individual rationality just stated implies that if the players are individually rational, the payoff vector must satisfy the inequality

$$x_i \geqslant v(\bar{i}) \quad (i = 1, 2, \ldots, n). \qquad (3.2)$$

Next, if the principle of group rationality is to be

served, we might expect that no subset of players should get less than they can get by joining a coalition. Formally speaking,

$$v(S) \leqslant \sum_{i \in S} x_i \text{ for all } S. \qquad (3.3)$$

In particular,

$$\sum_{i=1}^{n} x_i = v(N). \qquad (3.4)$$

Note that the last condition is an equality rather than an inequality. This is because while the grand coalition need not accept less than v(N), neither can they win more, since there are no more players to win it from.

As a first attempt to single out the payoff vectors that can be expected to serve as possible disbursements among rational players, let us consider all the vectors $\vec{x}$ which satisfy all three conditions (3.2), (3.3), (3.4). The set of such disbursements constitutes the *core* of the game.

Let us compute the core of our Left-Right game. Applying the inequalities (3.2) and (3.3), we must have, for any disbursement vector (x_1, x_2, x_3) which is in the core, the following inequalities satisfied:

$$x_1 \geqslant -2;\ x_2 \geqslant -4;\ x_3 \geqslant -4 \qquad (3.5)$$

$$x_2 + x_3 \geqslant 2;\ x_1 + x_3 \geqslant 4;\ x_1 + x_2 \geqslant 4. \qquad (3.6)$$

Now adding the last three of these inequalities, we obtain,

$$2(x_1 + x_2 + x_3) \geqslant 2 + 4 + 4 = 10 \qquad (3.7)$$

$$x_1 + x_2 + x_3 \geqslant 5. \qquad (3.8)$$

But the most that the three players can get jointly is $v(\overline{123}) = 0$ [cf. (2.6)]. Therefore inequalities (3.5) and (3.6) cannot be satisfied by the components of any disbursement vector which satisfies (3.8). In other words, the core is an empty set (cf. p. 17).

Similarly we see that the game defined by the characteristic function given by (2.9)–(2.12) (cf. p. 81) has an empty core. This is because each player can get 2 in a coalition with himself alone. All three cannot get 2 in a grand coalition, because the grand coalition can achieve a joint payoff of only 3. Since at least one player violates the principle of individual rationality by accepting a payoff of less than 2, and no disbursement vector satisfying (2.12) exists which can give 2 to each player, the core must be empty.

The situation looks different if we re-define the characteristic function of the Three-person Prisoner's Dilemma game so as to designate the value of the game to each subset of players as the very least the subset can get in a coalition (assuming that the counter-coalition will try to keep them to that minimum). Conceived in this way, the characteristic function is

$$v(\overline{1}) = -1;\ v(\overline{2}) = -1;\ v(\overline{3}) = -1 \qquad (3.9)$$

$$v(\overline{23}) = 0;\ v(\overline{12}) = 0;\ v(\overline{13}) = 0 \qquad (3.10)$$

$$v(\overline{123}) = 3. \qquad (3.11)$$

Now there are disbursement vectors which satisfy both individual and group rationality. One such vector is (1, 1, 1). We see that every player gets more than he could get by himself; each pair gets more than they could get in coalition; and all three get as much as they can get in this game.

There are, of course, other disbursement vectors in this game which satisfy these conditions. The vector (−1/2, 1/2, 3) is one. It may seem to be grossly unfair, especially to player 1; but that is another question. So far, we have demanded only that a disbursement vector satisfy both the principles of individual rationality and of collective rationality. We see that the game with the characteristic function given by (3.9)–(3.11) has such vectors. Its core is not empty.

It can be proved that all constant-sum N-person games

have empty cores. This makes the core unsuitable, at least in the context of constant-sum games, as a "prescription to rational players," in the sense of suggesting how they should divide the joint payoff (which is always the same regardless of outcome in a constant-sum game). We must therefore seek other bases for constructing "solutions" of such games.

We can drop the condition of group rationality for all the proper subsets of N and retain it only for the entire set N. This leaves intact the condition of individual rationality and the condition of group rationality for the entire set N. That is to say, we now impose the following conditions:

$$x_i \geqslant v(\bar{i}) \quad (i = 1, 2, \ldots, n) \qquad (3.12)$$

$$\sum_{i=1}^{n} x_i = v(N). \qquad (3.13)$$

Let us see why condition (3.3) (cf. p. 89) is the most reasonable one to drop. Of these three conditions (3.2), (3.3), and (3.4), the first is the most difficult to drop, because individual rationality is most difficult to dispense with. We have seen that in refusing to accept a payoff less than the value of the game to him, a player need not effect an agreement with anybody. It is therefore reasonable to suppose that "individual rationality" is an easy condition to satisfy. Of the other two conditions, it seems more difficult to drop the third than the second. This is because in forming a *grand coalition* the players do not face the possibility of other players luring members away (since there are no other players). In forming coalitions of *proper* subsets of N, on the other hand, the prospective players may have to face this problem. That is, players not in the prospective coalition may be offering some members of the coalition inducements to leave it. Thus, bargaining may ensue not only about how to divide v(N) but also about who is to be in coalition with whom. Under these conditions it may be more difficult to reach agreements, and consequently results may obtain which will not "satisfy everyone," which is just what happens when some players in a subset get less than they could obtain in a coalition by themselves. (They may not be able to form a coalition.)

A payoff vector which satisfies conditions (3.12) and (3.13) is called an *imputation*. Clearly the set of imputations of a game is always at least as large, in general larger, than the core of the game. In particular, games with empty cores (e.g., essential constant-sum games) always have imputations. The set of imputations, however, is too large to constitute a "solution" of the game. Here we have used "solution" in an intuitive, not in a precise sense. We should like a solution to satisfy some reasonable criteria about how payoffs should be disbursed among rational players. At the same time the disbursements which satisfy these criteria should be only a fraction of all possible disbursements, the smaller the fraction the better, from the point of view of a theory. In fact, if we could find a *unique* disbursement which would satisfy some reasonable criteria of "rationality," this would be the most significant result of the theory. Our problem, therefore, is to find constraints to be put on imputations so as to narrow down the set. We have seen that the constraint of group rationality (applied to every subset of players), which defines the core, is too strong for constant-sum games: it eliminates altogether the set of imputations which satisfy this constraint. In the following chapters, we shall examine some alternative constraints.

4. The Von Neumann-Morgenstern Solution

We may ask what meaning is to be assigned to the statement that one imputation is "better" than another. Clearly, any answer must implicitly answer the question "Better for whom?" In the Two-person zerosum game, of two distinct imputations, one must be better for one player while the other is better for the other. In non-zerosum games one imputation may be better for both players than another. These exhaust the possibilities for Two-person games. As the number of players increases, the number of possibilities rapidly increases. We can examine a pair of imputations to see whether one is better than another for any of the several subsets of players. This will certainly be the case if each member of the subset in question gets more in one imputation than he gets in another. If so, then the imputation which is "better" for the subset in question is said to *dominate* the other *via* the subset in question, provided the subset in question can actually get the larger amount by forming a coalition.

As an example, consider the two following imputations:

$$\vec{x}_1 = (1, 2, 3) \tag{4.1}$$

$$\vec{x}_2 = (2, 3, 1). \tag{4.2}$$

$\vec{x}_2$ is clearly better than $\vec{x}_1$ for players 1 and 2. Whether $\vec{x}_2$ dominates $\vec{x}_1$ via the set of players $\{1, 2\}$ depends on whether by forming a coalition $(\overline{12})$, players 1 and 2 can actually (jointly) get 5. This, in turn, depends on the value of the characteristic function associated with the coalition $(\overline{12})$. If $v(\overline{12}) \geqq 5$, then $\vec{x}_2$ dominates $\vec{x}_1$ via $(\overline{12})$.

We can now extend our definition of *domination* so that it will not depend on specific subsets of players. We shall say $\vec{x}_2$ dominates $\vec{x}_1$ if and only if there *exists* a subset of players who can jointly get (in a coalition) at least the sum of their payoffs in $\vec{x}_2$, and moreover if each of the members of this subset gets more in $\vec{x}_2$ than in $\vec{x}_1$.

N-person game theory would be immensely simplified if the domination relation were a weak ordering (cf. p. 29). Then one could rank order all possible imputations (the order of "equal" ones being arbitrary), select the imputation (or imputations) which dominate all others, and declare that these are the only possible outcomes of the N-person game among rational players. Unfortunately, except in the case of rather trivial games, this is not the case. It may happen that $\vec{x}_2$ dominates $\vec{x}_1$ (in the extended sense), while $\vec{x}_1$ does not dominate $\vec{x}_2$. Then $\vec{x}_2$ would be "higher" on our rank order of imputations. The case where neither $\vec{x}_2$ dominates $\vec{x}_1$ nor $\vec{x}_1$ dominates $\vec{x}_2$ presents no difficulty, since in that case $\vec{x}_1$ and $\vec{x}_2$ can be viewed as having the same "rank." It may also happen that two imputations dominate each other. These, too, can be assigned "equal rank." It may also happen, however, that $\vec{x}_1$ dominates $\vec{x}_2$, $\vec{x}_2$ dominates $\vec{x}_3$, and $\vec{x}_3$ dominates $\vec{x}_1$. Such instances violate the principle of transitivity (cf. p. 29) and preclude a meaningful ordering of the imputations with respect to the dominance relation.[10] The following is an example, assuming $v(\overline{23}) = v(\overline{13}) = v(\overline{12}) = 100$.

$$\vec{x}_1\text{: } (50, 50, 0) \tag{4.3}$$

$$\vec{x}_2\text{: } (0, 60, 40) \tag{4.4}$$

$$\vec{x}_3: \quad (15, 0, 85). \qquad (4.5)$$

We see that $\vec{x}_2$ dominates $\vec{x}_1$ via $(\overline{23})$; $\vec{x}_3$ dominates $\vec{x}_2$ via $(\overline{13})$, and $\vec{x}_1$ dominates $\vec{x}_3$ via $(\overline{12})$.

In spite of these ambiguities, the concept of a "solution" of an N-person game based on the idea of imputation domination has been proposed by Von Neumann and Morgenstern.

Consider a set of imputations $J = \{(x_1, x_2, \ldots, x_n)\}$ with the following properties:

1. No imputation in J dominates any other imputation in J ("internal stability").

2. If $\vec{x}$ is an imputation not in J (i.e., $\vec{x} \subset -J$), then there exists at least one imputation in J which dominates $\vec{x}$ ("external stability").

Such a set of imputations (which can with some justification be called a "stable set") constitutes a *solution* of the N-person game in the sense of Von Neumann and Morgenstern. (We shall also refer to such solutions as Von Neumann-Morgenstern solutions.) A game may have many solutions and particular imputations may belong to more than one solution; that is, the sets which constitute solutions are generally not disjoint.

The intuitive justification of this conception of solution stems from a feeling that the imputations in a solution are somehow more "stable" than those outside. If domination induces a shift (in the process of bargaining) from a dominated to a dominating imputation, there would be shifts "into" a solution. Moreover, since *within* a solution, no imputation dominates any other, each of the imputations in a solution might be expected to possess some sort of stability.

Here one must guard against a serious misconception. Recall that domination is not a "one way" relation. Two imputations may well dominate each other. Consequently the fact that every imputation outside a solution is dominated by some imputation in the solution by no means implies that the imputations in a solution are not dominated by imputations outside the solution. To be sure, we can single out just those imputations in a solution (or, for that matter from the entire set of imputations) which are not dominated by any imputations. This set is the *core* of the game (cf. p. 89). The core has a much stronger claim on "stability" than a Von Neumann-Morgenstern solution. Unfortunately, as we have already said, many games do not have cores (those sets turn out to be empty), and so the core does not provide a general basis for singling out the "stable" imputations.[11]

Even though the Von Neumann-Morgenstern solutions do not single out unique imputations as *the* outcomes of games played by rational players, this does not in itself make these concepts useless for a normative theory. Solutions which single out *sets* of outcomes rather than unique "outcomes" are common in mathematics. For example, the solution of a quadratic equation $ax^2 + bx + c = 0$ $(a \neq 0)$ comprises, in general, two values of x (in the realm of complex numbers), while the solution of the inequality $x^2 + y^2 < r^2$ (r: real) comprises an infinity of values, namely, all the points inside a circle with radius r around the origin as center.

Thus the mathematician is accustomed to "solutions" which merely narrow down the range of "outcomes" (to those which satisfy the imposed conditions) without necessarily narrowing them down to a single outcome. Nor is this notion of solution necessarily a crippling one for behavioral science (if N-person game theory is thought of as a possible paradigm for conflict situations involving more than two sets of interests). If the range of possible outcomes of such a game can be narrowed down, then the corresponding theory becomes "refutable"; that is, a result *not* in the range specified becomes a refutation or at least a challenge to the theory. Note that every substantive theory must provide for a possibility of being refuted, since, if every *possible* observation is consistent with the theory, the theory gives no new knowledge.

Let us see how the Von Neumann-Morgenstern theory fares in this respect when applied to a constant-sum Three-person game. Consider the Left-Right game described in Chapter 2. Its characteristic function, we have seen, is

$$v(\overline{1}) = -2;\ v(\overline{2}) = -4;\ v(\overline{3}) = -4 \qquad (4.6)$$

$$v(\overline{23}) = 2;\ v(\overline{13}) = 4;\ v(\overline{12}) = 4 \qquad (4.7)$$

$$v(\overline{123}) = 0;\ v(\overline{\emptyset}) = 0. \qquad (4.8)$$

An imputation in this case is any triple of payoffs where the sum is 0 and where each player gets at least $v(\overline{i})$, $(i = 1, 2, 3)$.

Consider the set J of imputation where $x_3 = -3$; $x_1 + x_2 = 3$ ($x_1 \geqslant 0$, $x_2 \geqslant 0$). We can see that no imputation in this set dominates any other imputation in the same set. This is because player 3's payoff being fixed, he does not prefer any other imputation; but players 1 and 2 cannot both prefer another imputation, since the sum of their payoffs remains constant; so that, if one gets more in the proposed imputation, the other must get less. Nor do all three players prefer any imputation in J to any other in J. Therefore the first condition to be satisfied by a "solution" is indeed satisfied; no imputation in the set dominates any other imputation in the set.

Now consider some imputation outside the set J. In such an imputation, player 3 must get either more or less than -3 (since *all* the imputations in which he gets -3 were included in J). Suppose player 3 gets more than -3 (say -2.8). Then, since this game is constant-sum, players 1 and 2 must get jointly less in this imputation than they get in J. Consequently there exists an imputation in J preferred by both 1 and 2 since they can always split their joint payoff so that each separately gets more than in the proposed imputation. In other words, there is always an imputation in J which dominates any imputation in −J (the complement of J with respect to the set of imputations) where player 3 gets more.

Suppose now player 3 gets less than −3, say −3.2 (he cannot get less than −4). Then 1 and 2 jointly get 3.2. Moreover, player 1 cannot get more than 7.2 (because player 2 must receive at least −4), and player 2 cannot get more than 5.2 (because player 1 must receive at least −2). Then player 3 prefers any imputation in J (where he gets −3 instead of −3.2). Moreover, no matter how players 1 and 2 split the 3.2 units (consistent with each getting no less than his value), there is *some* imputation in J which *one* of them prefers. If they split 3.2 equally, so that each gets 1.6, then player 1, say, can get more in the imputation (2, 1, −3) which is in J. If they do not split equally, then the player who gets less can certainly find a better imputation in J. It follows that there is some imputation in J which dominates any given imputation in −J. Thus, both conditions of "solution" are satisfied, and so every imputation in J belongs to a solution.

Can we extend J by adding other imputations to it and still have a solution? Clearly not, for we have seen that any imputation outside of J is dominated by some imputation in J, and these two cannot belong to the same solution by the definition of a solution. Thus, it appears that we have isolated a subset of imputations as a solution, which is what a theory of solutions purports to do.

Unfortunately for the theory, J is not the only solution of our game. Consider the set of imputations J′ in which player 3 receives the fixed amount −1.5, while players 1 and 2 divide the 1.5 between them. By exactly the same argument, *this* set of imputations can also be shown to be a solution. Can it be that *any* set of imputations in which player 3 receives a fixed amount, while players 1 and 2 split the complementary amount (in any proportion consistent with the characteristic function), turns out to be a solution? We can show by a counter-example that this is not so.

Consider the set of imputations J″, in which player 1 receives 3 while players 2 and 3 receive q and $-3-q$ respectively. To be sure, within such a set no imputation dominates any other, since x_3 is fixed, and the interests of players 1 and 2 are always opposed. However, the imputation $(-2, 1, 1)$ is not dominated by any imputation in J″; for while player 1 certainly prefers 3 to -2, in no imputation of J″ can either player 2 or player 3 receive more than 1, since this would give the other less than his value. Thus not every set of imputations in which one player receives a fixed amount, while the others split what is left in any way consistent with individual rationality, constitutes a solution.

Nevertheless the Von Neumann-Morgenstern solution theory applied to certain games is irrefutable and therefore devoid of theoretical leverage as it stands. We shall illustrate by examining the Three-person constant-sum game in its normalized form:

$$v(\bar{i}) = 0 \quad (i = 1, 2, 3) \tag{4.9}$$

$$v(\overline{ij}) = 1 \quad (i, j = 1, 2, 3) \tag{4.10}$$

$$v(\overline{123}) = 1. \tag{4.11}$$

In other words, the three players are to divide up one unit by "majority vote," and no player is obligated to pay out of his own pocket. One solution of this game is the set consisting of three imputations:

$$(1/2, 1/2, 0); (1/2, 0, 1/2); (0, 1/2, 1/2). \tag{4.12}$$

Further, any set of imputations (x_1, x_2, c) where $0 \leqslant c < 1/2$ and $x_1 + x_2 = 1 - c$ constitute a solution.

Likewise the analogous sets (x_1, c, x_2) and (c, x_2, x_3) constitute solutions. In other words, the solutions can be enumerated as follows:

1. The set of imputations where one player gets nothing, and the other two split the unit equally.

2. Each set of imputations where one player gets a fixed amount less than 1/2, while the other two split what remains in any manner whatsoever (neither getting a negative payoff).[12]

Clearly there is an infinite number of solutions (indeed, a triply-infinite number of them) since x_1, x_2, and c vary over continuous intervals.

The infinity of solutions and imputations, however, is not what deprives the Von Neumann-Morgenstern solution of its theoretical leverage. There would be considerable theoretical leverage left if it could be shown that some imputations were *not* members of any solution. Then the theory would declare some imputations "impossible" (i.e., incompatible with the hypothesis that a realizable imputation must be in some solution), and the theory could be put to a test. Unfortunately, this is not the case. *Every* imputation of this game is in *some* solution. To see this, consider any imputation (a_1, a_2, a_3), where, of course, $a_i \geqslant 0$, and $\sum_{i=1}^{3} a_i = 1$. Then at least one of the a_i must be less than 1/2. Let it be a_3, and call it c. Then the set of imputations (a_1, a_2, c) is a solution which contains the imputation proposed.

If every imputation is in some solution, then no matter what outcome we observe (consistent with the rules of the game), we cannot say that a prediction of the solution theory has been violated. There is, however, one rather weak justification of the theory. Suppose we observe that when a certain set of players play a Three-person constant-sum game many times, one of the players (always the same one) gets a fixed amount c ($0 \leqslant c < 1/2$), while the other two players divide up the remainder among them, not necessarily in the same proportion each time. In this case one could say that the three players have singled out *a* solution and have adhered to it. Perhaps in these circumstances our attention will be directed to the characteristics of the player who gets the fixed amount. He may be a privileged person (in case he gets more than 1/3) to whom the other two players feel they must give his due; or he may be a person who is willing to settle for something less than his "equitable" expectation (if he gets less than 1/3), just to avoid the trouble of bargaining. Or he may be a pariah (in case he gets 0) whom the others deliberately exclude from the game, regardless of what he offers either of them in return for joining in a coalition. (Note that to identify this special person we must observe in the course of iterated plays that the other two split between them in *different* proportions, or else we cannot tell which of possible two or three players who get less than 1/2 is the man with the fixed payoff.)

Also, if we observed that two players split equally, while the third gets nothing (not necessarily always the same player), then we could likewise say that a particular solution, namely, the symmetric one (see Note 12) has been "singled out." Note, however, that if one player always got more than 1/2 while the other two split in *different* proportions in repeated plays, we could not say that a solution has been singled out. For example, the set of imputations $J = \{(x_1, x_2, 3/4)\}$ where $x_1 + x_2 = 1/4$ do not constitute a solution. To see this, observe that no imputation in this set dominates (1/4, 1/4, 1/2), since neither player 1 nor player 2 can get more than 1/4 in any imputation of J.

Let us see how the solution appears in our original (un-normalized) game. To see this we "undo" the normalization. That is, we first change the unit of payoff so that an imputation is given by

$$x_1 + x_2 + x_3 = 10 \tag{4.13}$$

$$x_i \geqslant 0 \quad (i = 1, 2, 3). \tag{4.14}$$

Then we take away from each player the constant sum we had given him to normalize the game, namely 2 from player 1, and 4 from players 2 and 3 respectively. In this way the imputation (x_1, x_2, x_3) of the normalized game becomes the imputation $(10x_1 - 2, 10x_2 - 4, 10x_3 - 4)$ of the original game. In particular, the solution given by (4.12) becomes

$$(3, 1, -4); (3, -4, 1); (-2, 1, 1). \qquad (4.15)$$

Solutions of the form (c, x_2, x_3), (x_1, c, x_3), and (x_1, x_2, c) become of the form (c_1, x_2, x_3), (x_1, c_2, x_3), (x_1, x_2, c_3), where $-2 \leqslant c_1 < 3$; $-4 \leqslant c_2 < 1$; $-4 \leqslant c_3 < 1$; $x_1 \geqslant -2$, $x_2 \geqslant -4$, $x_3 \geqslant -4$; $x_i + x_j = -c_k$ for distinct i, j, k = 1, 2, 3.

The Von Neumann-Morgenstern solutions of Three-person games can be conveniently visualized in geometrical representation. Let us return to the normalized game. A payoff vector being a triple of numbers can be represented by a point (x, y, z) in ordinary three-dimensional space (cf. p. 36). The imputations of the game are all the points (x, y, z), in which

$$x + y + z = 1 \qquad (4.16)$$

$$x \geqslant 0; y \geqslant 0; z \geqslant 0. \qquad (4.17)$$

These are the points on the boundaries and inside the triangle intercepted by the plane $x + y + z = 1$, and the three coordinate planes. This triangle is shown in Figure 9, with its vertices labeled by their respective coordinate vectors. The solution $(1/2, 1/2, 0)$, $(1/2, 0, 1/2)$. $(0, 1/2, 1/2)$ appears as three mid-points on the sides of the triangle. The other solutions are shown as families of straight lines parallel to the respective sides of the triangle, including the sides themselves.

The Von Neumann-Morgenstern solutions of Three-person non-constant-sum games are considerably more complicated. They will be briefly mentioned in Chapter 12. A complete discussion of them is beyond the scope of this book. By way of example, typical solutions of Three-

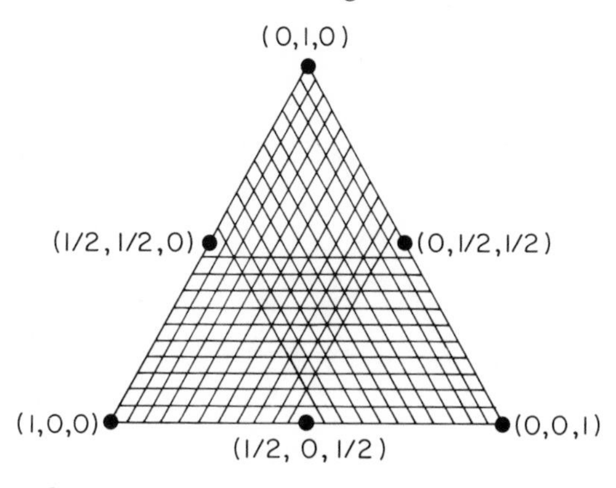

FIG. 9. Solutions of the Three-person constant-sum game. The triangle is the imputation simplex; three midpoints of the sides of the triangle constitute the symmetric solution (see Note 12). Each line parallel to one of the sides is a discriminatory solution. Note that the entire imputation simplex is "covered" by solutions, so that every imputation is in some solution. This is not the case in other games (cf. Figures 10 and 11).

person non-constant-sum games are shown in Figures 10 and 11.

If $n > 3$, solutions become even more complex sets.

As we have said, the singling out of a solution may be manifested in the circumstance that a set of players, playing the game in question, consistently arrive at one of the imputations which belong to a set that constitutes a particular solution. This may reflect some standards or "social norms" which govern the behavior of the players. Of course, a social norm may narrow down the set of realized imputations to a much smaller set than a solution. We may observe, for example, that the players recruited from a given population, when playing the Three-person constant-sum game, always split the joint payoff equally. This pattern of behavior does not single out a solution, since the imputation (1/3, 1/3, 1/3) does not by itself constitute a solution of this game in the Von Neumann-Morgen-

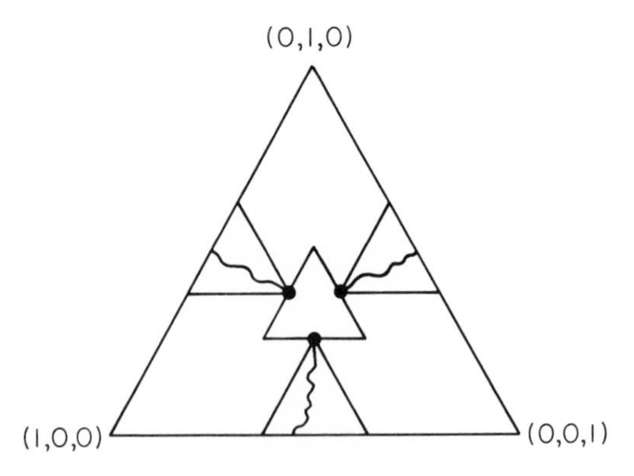

FIG. 10. (After Von Neumann and Morgenstern, *Theory of Games and Economic Behavior,* 2nd. ed., Figure 82). A typical solution of a normalized Three-person non-constant-sum game. The large triangle is the imputation simplex. The small central triangle is determined by the specific characteristic function. The particular solution is the set of all points on the curves including the points on the sides of the small triangle.

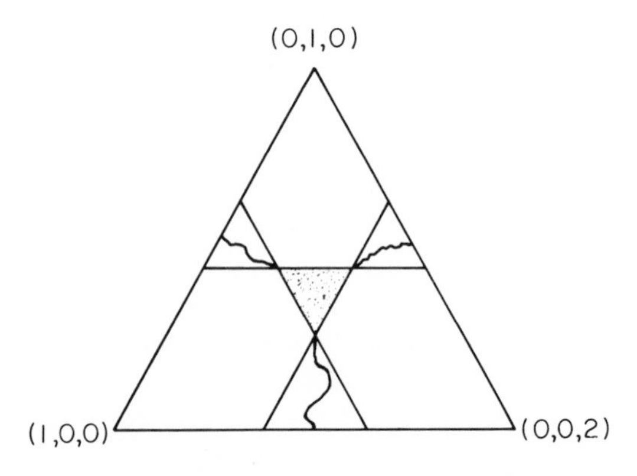

FIG. 11. (After Von Neumann and Morgenstern, *Theory of Games and Economic Behavior,* 2nd. ed., Figure 86). Another typical solution of a normalized Three-person non-constant-sum game: all the imputations in the shaded triangle and on the curves going from its vertices to the sides of the large (imputation simplex) triangle.

stern sense. To explain the results, one could easily invoke some social norm (e.g., equity) without recourse to the Von Neumann-Morgenstern theory of solutions. In the next chapter we shall examine another approach which does single out a unique imputation of an N-person game by invoking the a priori expectations of the players with regard to their respective bargaining positions.

5. *The Shapley Value*

Consider a Three-person non-constant-sum game specified by the following characteristic function

$$v(\overline{1}) = 0; v(\overline{2}) = 1; v(\overline{3}) = 1 \qquad (5.1)$$

$$v(\overline{23}) = 3; v(\overline{13}) = 4; v(\overline{12}) = 5 \qquad (5.2)$$

$$v(\overline{123}) = 16. \qquad (5.3)$$

Suppose a rational player considers his prospects as player 1 in this game. He knows that the other two players are also rational and that both coalitions and side payments will be allowed. He can expect that each coalition can count on getting at least as much as the characteristic function awards to it. But then the question arises as to how these joint payoffs are to be apportioned among the members of the coalitions. Assuming the players to be *collectively* rational, player 1 can foresee that the grand coalition will eventually form, since all three jointly can get more by working together than in any partition into separate coalitions. The question, that player 1 is considering, is how to bargain as the grand coalition is shaping up.

"Suppose," he muses, "player 2 approaches me with the proposition to form a coalition with him. His proposition might be something of this sort:

'I, player 2, can get 1 unit playing alone. Together with you, player 1, we can get 5. Shall we join?' "

To this, player 1 is planning to reply as follows:

"Since our joint gain will exceed your own prospect by 4 units, you can afford to give me up to 4 units out of our joint gain of 5."

Player 1 cannot, of course, expect that player 2 will

immediately agree to this. But he makes a mental note of this limit of his expectation should this situation arise. Next he contemplates the possibility that player 3 will approach him. To him player 1 can say:

"Your prospect, playing alone, is 1. Together we get 4. Therefore you can afford to give me up to 3."

Next, player 1 supposes that players 2 and 3 have already formed a coalition. What should he demand of them, if they approach him with a proposition to join as the third (and last) member of the grand coalition? He can say to them:

"You two can get 3. With me, we get 16. Therefore I am worth 13 to you as the third member of the coalition. Give me 13."

We see, then, that player 1's bargaining power depends on which coalition he is about to join.

Let us suppose that the grand coalition will eventually form by growing from one to two to three members. Without any specific information on the order in which the members will join, let us suppose that all orders are equally likely. Thus the following orders of joining may arise with the probabilities as noted.

(1, 2, 3) with probability 1/6
(2, 3, 1) "
(3, 1, 2) "
(3, 2, 1) "
(2, 1, 3) "
(1, 3, 2) "

Thus, in two of the six situations, player 1, being the first member, was not solicited by anyone. He, in fact, joined the empty coalition. By joining it, he brought an increment of $v(1) - v(\varnothing)$ into the (empty) coalition. In our case this happens to be zero. Therefore, with probability $1/6 + 1/6 = 1/3$, he can expect to get 0.

In one of the situations he was solicited by player 2. Here, with probability 1/6, he expects to get 4. In one he was solicited by player 3. Here, with probability 1/6, he expects to get 3. In the remaining two situations, he was solicited by the coalition $(\overline{23})$. Here, with probability 1/3, he expects to get 13. His expected prospect, therefore, is

$$1/3 \times 0 + 1/6 \times 4 + 1/6 \times 3 + 1/3 \times 13 = 11/2. \tag{5.4}$$

Calculating the expected prospects of player 2 in exactly the same way, we obtain

$$1/3 \times 1 + 1/6 \times (5 - 0) + 1/6(3 - 1) + 1/3(16 - 4) = 11/2; \tag{5.5}$$

and for player 3:

$$1/3 \times 1 + 1/6(3 - 1) + 1/6(4 - 0) + 1/3(16 - 5) = 5. \tag{5.6}$$

These expectations will now be called the *Shapley value* of the game to players 1, 2, and 3 respectively. Note that the three values add up to the value of the game to the grand coalition. Let us see whether this would be true in the general case. For simplicity, we shall investigate only the Three-person game.

The game is given, as usual, by its characteristic function $v(S)$, where S ranges over all the subsets of N. The prospective gains to player 1 are the differences $v(S) - v(S - \{1\})$, where S ranges over all the coalitions of which player 1 can be a member. Player 1's prospect is the weighted average of these differences, the weights being the respective *probabilities* that player 1 joins the coalition $(S - \{1\})$ as a new member. Note that player 1 can be a member of four coalitions, namely $(\overline{1})$, $(\overline{12})$, $(\overline{13})$, and $(\overline{123})$. The weights in this average are the probabilities of occurrence of the relevant situations. Hence, designating the values by $\varphi_i (i = 1, 2, 3)$, we have:

$$\varphi_1 = 1/3[v(\overline{123}) - v(\overline{23})] + 1/6[v(\overline{12}) - v(\bar{2})] + 1/6[v(\overline{13}) - v(\bar{3})] + 1/3[v(\bar{1}) - v(\emptyset)] \quad (5.7)$$

$$\varphi_2 = 1/3[v(\overline{123}) - v(\overline{13})] + 1/6[v(\overline{12}) - v(\bar{1})] + 1/6[v(\overline{23}) - v(\bar{3})] + 1/3[v(\bar{2}) - v(\emptyset)] \quad (5.8)$$

$$\varphi_3 = 1/3[v(\overline{123}) - v(\overline{12})] + 1/6[v(\overline{13}) - v(\bar{1})] + 1/6[v(\overline{23}) - v(\bar{2})] + 1/3[v(\bar{3}) - v(\emptyset)]. \quad (5.9)$$

Adding and simplifying, we get

$$\varphi_1 + \varphi_2 + \varphi_3 = v(\overline{123}). \quad (5.10)$$

For the general N-person game, it can be shown that

$$\varphi_i = \sum_{S \subseteq N} \frac{(s-1)!(n-s)!}{n!} [v(S) - v(S - \{i\})]. \quad (5.11)$$

The summation ranges over all the subsets S of N, and $s = |S|$ (the number of members of S). The proof that

$$\sum_{j=1}^{n} \varphi_i = v(N) \quad (5.12)$$

is straightforward.

The rationale we have given for the Shapley value is in terms of the bargaining power which each player imagines he possesses. This power (as estimated by the player in question) is based on what his joining each coalition contributes to that coalition. The rationale can be given also in another way, namely by means of an axiomatization. We demand that any n-tuple of payoffs $(x_1, x_2, \ldots, x_n)$, for which a claim is made as a "reasonable" outcome of the N-person game, satisfy three axioms. The first two of these are:

1. The vector $(x_1, x_2, \ldots, x_n)$ should depend only on the characteristic function of the game, not on any inherent properties of the players, in particular not on the way the players are labeled.

2. We must have

$$\sum_{i=1}^{n} \varphi_i = v(N). \quad (5.13)$$

Before we state the third condition, we must explain what is meant by a game which is a composite of two games.

Suppose we have two games, G_m and G_n, one with a set of players M, the other with a set of players N. The two sets of players may or may not be disjoint. Consider the union of the two sets of players $M \cup N$ and a subset S of this union (i.e., $S \subseteq M \cup N$). Let v( ) be the characteristic function of G_m and w( ) the characteristic function of G_n. Then we can write

$$v(S) = v(M \cap S);\ w(S) = w(N \cap S). \qquad (5.14)$$

This is because those and only those players of S who are in $M \cap S$ are "affected" by the characteristic function of G_m, i.e., can expect the minimum joint payoff accruing to them in that game. Similarly, those and only those players in S who are in game G_n are affected by its characteristic function. Now the composite of the games G_m and G_n is so defined that its characteristic function u(S) is the sum of the respective characteristic functions:

$$u(S) = v(S) + w(S). \qquad (5.15)$$

Let us now express the dependence of φ on the characteristic function of a game by the usual functional notation. Thus, $\varphi_i(v)$ will be the i-th component of the m-vector if G_m is played, and $\varphi_i(w)$ the i-th component of the n-vector if game G_n is played (assuming that player i is a player in both games). Finally $\varphi_i(v + w)$ will be the designation of the i-th component of the corresponding vector if the composite game G_{m+n} is played. The third axiom states that

3. $$\varphi_i(v + w) = \varphi_i(v) + \varphi_i(w). \qquad (5.16)$$

That is to say, if the i-th player is in both games, the minimal expected payoff accruing to him in the composite game shall be the sum of the minimal expected payoffs accruing to him in the separate games. Note that if player i is in only one of the games, the payoff accruing to him in the other is zero; so this means that he gets the same payoff in the composite game as in the original game in which he was a player.

Luce and Raiffa (*Games and Decisions*, p. 248) point out that this third assumption is not nearly as easy to accept as the first two.

"For," they write, "although v + w is a game composed from v and w, we cannot in general expect it to be played as if it were the two separate games. It will have its own structure which will determine a set of equilibrium outcomes which may be very different from those for v and w. Therefore one might well argue that its a priori value should not necessarily be the sum of the values of the component games. This strikes us as a flaw in the concept of value, but we have no alternative to suggest."

In referring to the structure of the composite game and to its equilibrium outcomes, Luce and Raiffa are looking beyond the characteristic function description of the game, looking at it with greater "resolving power," as it were. We have already seen how the characteristic function formulation discards a great deal of information about the game. We shall take this information into account in Chapters 10 and 11. It is, however, worthwhile to pursue the consequences of the assumption that can be made on the basis of the characteristic function only.

Let us see what the Shapley value formulation has accomplished. The most important and gratifying consequence of the three axioms is that the payoff vector which satisfies all three is *unique*. Next, it has some very desirable properties. To illustrate, let us "solve" the Three-person constant-sum game by deriving its Shapley value payoff vector. Using the normalized version of this game, we have

$$v(\overline{1}) = v(\overline{2}) = v(\overline{3}) = 0;$$
$$v(\overline{12}) = v(\overline{13}) = v(\overline{23}) = v(\overline{123}) = 1. \quad (5.17)$$

We can see from the symmetry of this game that we shall have

$$\varphi_1 = \varphi_2 = \varphi_3 = 1/3. \tag{5.18}$$

The Shapley value solution, therefore, awards each player of this game an equal amount. To be sure, it would not appear so in a non-normalized form of the game. Suppose, for example, our constant-sum game were represented by the following characteristic function:

$$v(\overline{1}) = 1;\ v(\overline{2}) = 0;\ v(\overline{3}) = -2 \tag{5.19}$$

$$v(\overline{23}) = 2;\ v(\overline{13}) = 3;\ v(\overline{12}) = 5 \tag{5.20}$$

$$v(\overline{123}) = 3. \tag{5.21}$$

Applying formulas (5.7)–(5.9) we obtain

$$\varphi_1 = 1/3(3-2) + 1/6(5-0) + 1/6(3+2) + 1/3(1-0) = 14/6 \tag{5.22}$$

$$\varphi_2 = 1/3(3-3) + 1/6(5-1) + 1/6(2+2) + 0 = 8/6 \tag{5.23}$$

$$\varphi_3 = 1/3(3-5) + 1/6(3-1) + 1/6(2-0) + 1/3(-2) = -4/6. \tag{5.24}$$

To normalize this game, we take from player 1 a "fee" of 1 unit and pay player 3 a bonus of 2 units. Since this makes $v(\overline{123}) = 4$, we must change the unit of payoff by dividing by four. Consequently, to "de-normalize" the game so as to obtain the original game, we must multiply the units by 4. This gives everyone 4/3. Then we pay back 1 to player 1 and take back 2 from player 3. This leaves player 1 with $4/3 + 3/3 = 7/3 = 14/6$, and player 3 with $4/3 - 6/3 = -2/3 = -4/6$. Player 2's payoff remains unchanged.

We see, then, that in perfectly symmetric games, where each player is in exactly the same position vis-à-vis others, the Shapley value solution awards the same payoff to each; in games where the positions of the players differ, the Shapley value reflects this difference.

Observe, for example, the positions of the players in the "non-symmetric" constant-sum game. From the characteristic function, we see that player 1's position is strongest and player 3's position is weakest (comparing the guaranteed payoff to each). This is also reflected in the differences in their Shapley value payoffs.

It seems, therefore, that the Shapley value solution, besides determining a unique disbursement of the payoffs solely by the characteristic function of the game, has built into it a certain equity principle. This solution might therefore be a strong contender for the status of a "normative" solution, i.e., one which "rational players" ought to accept. Its weakness is precisely in that it derives entirely from the characteristic function of the game and not from what is "behind" the characteristic function, i.e., the strategic structure of the game itself rather than the bargaining positions of the players in the process of coalition formation.

This aspect of the Shapley value solution can be most directly understood when we apply it in a simpler situation, namely the Two-person cooperative game. This we shall do in Chapter 10. In the interest of gaining perspective on the N-person game, we shall also examine the meaning of the other concepts so far discussed; namely, the characteristic function, normalization, and the Von Neumann-Morgenstern solution, as they appear when projected onto the simpler context of the Two-person game. This comparison will reveal in a simple context just what is lost when we pass from the normal form to the characteristic function of the game. In the final chapter of Part I we shall describe a model where this "loss" is restored.

6. *The Bargaining Set*

So far we have raised questions concerning only the disbursement of payoffs among the players of an N-person game, not questions concerning the coalitions which will form. The reason is that in every concept we have used so far there is an implication either that the grand coalition *has* formed or, at least, that the outcome is not inconsistent with the formation of a grand coalition. This idea is embodied in the definition of the imputation, which has been involved in all the concepts so far discussed, namely the Von Neumann-Morgenstern solution, the core, and the Shapley value. Recall that, if the outcome is an imputation, the players get jointly exactly what they would get jointly in the grand coalition. If the players get more in the grand coalition than they can jointly get in any other coalition structure, then obviously, to get the full amount, they must join in the grand coalition. At any rate, the assumption that the disbursement of the payoffs is an imputation is consistent with the assumption that the players have joined in the grand coalition.

In other words, the use of the imputation as the basis of a "reasonable outcome" implies, or at least does not contradict, that the principle of group rationality applies to the whole set of players.

Recall also that the use of the core of the game as the point of departure implies group rationality not only for the whole set but also for every subset of players. The fact that for some games the core is empty erodes the ground under this assumption. For, if *some* outcome obtains in a game with an empty core, it follows that the principle of group rationality has been violated for at least one subset of players. Once this principle is shaken, we are tempted to drop it also for the whole group. Admittedly, it is easier to drop "group rationality" for proper subsets of N than for the whole set, as has been argued above (p. 91). Let us nevertheless drop the principle of group rationality (of the n players) altogether. Then the outcomes are not confined to the imputations. The players can now get jointly less than accrues to them in the grand coalition. We shall, however, keep the principle of individual rationality, for without it the very raison d'être of game theory will be destroyed.

Once we drop the principle of group rationality applied to the whole set, the question concerning which coalitions will form becomes acute, for we can no longer assume that the grand coalition has formed. Actually the question of which coalitions will form is proper also in the context of constant-sum games (which, recall, have empty cores), since different coalition structures are consistent with the imputation being an outcome. In fact, it is naturally assumed that in the constant-sum game the grand coalition does *not* form. As an example, consider the essential Three-person constant-sum game. Its (normalized) characteristic function is

$$v(\overline{1}) = v(\overline{2}) = v(\overline{3}) = 0 \tag{6.1}$$

$$v(\overline{12}) = v(\overline{13}) = v(\overline{23}) = 1 \tag{6.2}$$

$$v(\overline{123}) = 1. \tag{6.3}$$

Once a coalition of two players has formed, there is no inducement for the two in the coalition to invite the third one in. Since the two already get as much as all three can get, the third player can bring nothing into the coalition.

Even if the three can get more than any two, the core may still be empty. This means, in effect, that if the grand coalition forms there will always be disgruntled players in it, i.e., players that can get jointly more if they go off by themselves than if they stay in the grand coalition. Under these circumstances it may happen that the grand coalition will not form or, if formed, will not last, and we shall then get outcomes which violate group rationality.

Therefore in this and the following chapters we shall not assume group rationality except for coalitions which actually form. That is, for a coalition to form, it will be necessary that its members get jointly at least the value of the game to the coalition; but we shall not require that any subset of players, whose members belong to *different* coalitions shall get as much as they could get in a single coalition. We shall assume that some coalitions have actually formed, satisfying only the condition stated. On the basis of this assumption we shall examine the "stability" of the coalition structure.

The point of departure is the so-called *individually rational payoff configuration.* We imagine that the players have joined several coalitions $B_1, B_2, \ldots, B_m$. If x_i is the payoff which player i gets as a member of a coalition, and $v(\bar{i})$ is the value of the game to him given by the characteristic function, we must, in view of individual rationality, have

$$x_i \geqslant v(\bar{i}). \tag{6.4}$$

Next, we can expect that the members of a coalition *jointly* will get the value of the game to the coalition. Hence

$$\sum_{i \in B_j} x_i = v(B_j). \tag{6.5}$$

The *individually rational payoff configuration* will now be defined in terms of any coalition structure (B_1,

$B_2, \ldots, B_m$) and a payoff vector satisfying (6.4, 6.5). We shall denote this configuration by $(\vec{x}, \mathcal{B})$. Here $\mathcal{B}$ denotes the set of sets $(B_1, B_2, \ldots, B_m)$ (cf. p. 22).

Consider now two players, k and ℓ, belonging to the same coalition B_j. Each player pursues his self-interest. He is therefore asking himself whether he can do better than he does. He might conceivably do better in another coalition. There are three ways of changing the coalition to which he belongs, namely

(a) by enticing additional players to join it;

(b) by excluding members from it;

(c) by both enticing other members to join it and by excluding members from it.

For the time being, let us ignore (a) and confine our attention to (b) and (c); in other words, k is contemplating forming some other coalition C (distinct from B_j, to which he presently belongs), from which at least one of his partners in B_j will be excluded. He contemplates all such coalitions in which he, player k, is included but from which player ℓ is excluded. Let coalition C be a member of this set.

Suppose player k notes that if coalition C is formed, each of its prospective members could get more than they are presently getting. In symbols, there exists a c-tuple of numbers $(y_1, y_2, \ldots, y_c)$ where $c = |C|$, such that

$$\sum_{i \in C} y_i = v(C) \tag{6.6}$$

and

$$y_i > x_i, \quad i \in C. \tag{6.7}$$

If this is the case, we shall say that player k has an *objection* against player ℓ. He can formulate this objection as follows:

"Why should I stick to the present coalition B_j with you, ℓ, a member of it, when c of us can form another coalition C (including me but excluding you) in which each of us can get more than we are now getting?"

Suppose now player k confronts player ℓ with this objection. Does player ℓ have recourse to a counter-argument? This depends. He, in turn, can consider all the possible coalitions in which he, ℓ, is included but from which k is excluded. Suppose among these he finds a coalition D in which the members (including him but excluding k) can each get more than they are presently getting. Now some of the players may be members of both the set C and the set D; in other words, of the intersection $C \cap D$. Then if both k and ℓ make their respective proposals of forming coalition C (k's proposal) and coalition D (ℓ's proposal), then the players in the intersection $C \cap D$ are presented with a choice between belonging to coalition C or to coalition D. Let us suppose that it is to each of these players' advantage to belong to D (i.e., each of them gets more in D). Then we shall say that player ℓ has a counter-objection to player k. Player ℓ can now say to k:

"If you propose coalition C, from which I am excluded, I can propose coalition D, from which you are excluded. The players who have to choose between the two coalitions (being prospective members of either) will choose my proposal rather than yours, since they have more to gain thereby."

In symbols, if $\mathfrak{J}_{\ell,k}$ is the set of all subsets of N which includes ℓ but not k, ℓ has a counter-objection to k if there exists a set D such that $D \in \mathfrak{J}_{\ell,k}$ and $z = (z_1, z_2, \ldots, z_d)$, $(d = |D|)$, $z_i \in D$, $(i = 1, 2, \ldots, d)$, and such that:

$$\sum_{i=1}^{d} z_i = v(D) \tag{6.8}$$

$$z_i \geqslant x_i \tag{6.9}$$

$$z_i \geqslant y_i, \quad i \in C \cap D. \tag{6.10}$$

It may happen that ℓ has no counter-objection against k's objection. Then we shall say that k has a *justified* objection against ℓ.

We are now ready to define the *bargaining set* $\mathfrak{M}$. It is the set of all individually rational payoff configurations in which no player has a justified objection against any other member of the same coalition. The elements of this set will be denoted by payoff disbursement vectors followed by a designation of a coalition structure. For example, in the Divide-the-Dollar game to be presently examined $(50, 50, 0; \overline{12}, \overline{3})$ is an element of $\mathfrak{M}$, as will appear below.

The Divide-the-Dollar game is a Three-person constant-sum game with $v(\overline{12}) = v(\overline{13}) = v(\overline{23}) = v(\overline{123}) = 100$; $v(\overline{i}) = 0$ $(i = 1, 2, 3)$. Suppose coalition $(\overline{12})$ forms. We need to examine each imputation with reference to the possible justified objections which 1 and 2 may have against each other.

Consider first the imputation (50, 50, 0). Player 1 has an objection against player 2, since he can say, "If I form a coalition with 3, from which you are excluded, I can offer 40 to 3, taking 60 for myself." However, the objection is not justified, since 2 has a counter-objection: "I can offer 45 to 3, taking 55 for myself." Whatever 1 can offer 3 up to 50 exclusive, 2 can better the offer; if 1 offers 50, 2 can match this offer. Hence the individually rational payoff configuration $(50, 50, 0; \overline{12}, \overline{3})$ is in the bargaining set of the Divide-the-Dollar game.

On the other hand, an imputation like $(60, 40, 0; \overline{12}, \overline{3})$ is not in the bargaining set, since 2 can offer 55 to 3 and come out better in a coalition with 3, while 1 cannot match this offer to 3. Therefore 2 has a *justifiable* objection against 1. By our definition, the individually rational payoff configuration $(60, 40, 0; \overline{12}, \overline{3})$ is not in the bargaining set.

Indeed, if two players form a coalition against the third, then the only individually rational payoff configurations are those where the players in the coalition get equal shares, while the third player gets nothing. These configurations correspond to the so-called *strong* solution proposed by W. Vickrey.[13] They correspond also to the symmetric solution of the Three-person constant-sum game (see Note 12).

If each player is in coalition only with himself, he has no player to have an objection against, and so the only individually rational payoff configuration is the trivial one $(0, 0, 0; \overline{1}, \overline{2}, \overline{3})$. Note that this configuration does not correspond to an imputation.

It remains to examine the individually rational configurations corresponding to the grand coalition. Clearly, they must all be imputations, since the coalition must get its value. Consider an imputation (a, b, c) where $a > b \geqslant c$. Then $b < 50$, $c < 50$. Now player 2 has a justified objection against player 1, for he can offer player 3 up to $100 - b$ in a coalition that excludes player 1. This is more than c and also more than $100 - a$, which is the most that player 1 can offer player 3 in a coalition that excludes player 2. Therefore the payoff configuration $(a, b, c; \overline{123})$ is not stable. Similarly, it is easy to see that if $a = b > c$, player 3 has a justified objection against either player 1 or player 2. Since the labeling of the players is arbitrary, we see that no payoff configuration associated with the grand coalition is stable if one player gets more than another. Moreover, the equally divided imputation is clearly stable.

In summary, the stable payoff configurations of the Divide-the-Dollar game are the following:

$$\begin{aligned}
&(0, 0, 0; \overline{1}, \overline{2}, \overline{3});\\
&(50, 50, 0; \overline{12}, \overline{3});\\
&(50, 0, 50; \overline{13}, \overline{2});\\
&(0, 50, 50; \overline{1}, \overline{23});\\
&(33\ 1/3, 33\ 1/3, 33\ 1/3; \overline{123}).
\end{aligned} \tag{6.11}$$

These, then, constitute the bargaining set of this game.[14]

Consider now a Three-person game (non-constant-sum) given by the generalized (non-super-additive) characteristic function

$$v(\bar{i}) = 0 \quad (i = 1, 2, 3) \qquad (6.12)$$

$$v(\overline{12}) = 60; v(\overline{13}) = 70; v(\overline{23}) = 90 \qquad (6.13)$$

$$v(\overline{123}) = 0. \qquad (6.14)$$

Suppose coalition ($\overline{23}$) forms and 3 proposes the split 52.5–37.5 of the 90 units, himself getting the larger share. His reasoning is approximately as follows:

"If you, 2, go into partnership with 1, the two of you get only 60. If you go with me, we get 90. Therefore, I am worth more as a partner to you than 1. If you split even with 1, you get 30. If you were to split even with me, you would get 45, a gain of 15. You ought to give me one half of that gain, and so be satisfied with 37.5."

Player 2 has now a justified objection against 3. He can offer 1 up to 22.5 and come out ahead, while 3 can offer 1 only up to 17.5.

On the other hand, the configuration $(0, 40, 50; \bar{1}, \overline{23})$, is in the bargaining set, since each offer made to player 1 by either player 2 or player 3 can be matched (in general, topped) by the other.

Similarly, it can be seen that if coalition ($\overline{12}$) forms, the disbursement (20, 40, 0) is the only "stable one"; if coalition ($\overline{13}$) forms, then (20, 0, 50) is the only stable disbursement.

In summary, the nontrivial configurations of the bargaining set of the game are

$$(20, 40, 0; \overline{12}, \bar{3}); (20, 0, 50; \overline{13}, \bar{2}); (0, 40, 50; \bar{1}, \overline{23}). \qquad (6.15)$$

Can we say anything about the coalitions which will actually form? Unfortunately not. Assuming that one of the outcomes of the bargaining sets will obtain, we see that no player has a preference between the two others as a coalition partner. Each gets the same payoff regardless of whom he joins with. One might think that player 1 might prefer player 3 to player 2 in the hope of getting a larger side payment from the former, who gets more. However, it will not do to consider the possible side payments which the players may offer to gain a partner, because this would result either in departing from the individually rational payoff configurations in the bargaining set, or in some player having a justified objection against his partner.

It turns out, then, that in the case considered, the notion of the bargaining set did enable us to single out a unique outcome to associate with each coalition structure, but it told us nothing about which coalition structure we could reasonably expect from rational players.

It is natural to suppose that extra-game-theoretical considerations govern the formation of coalitions, e.g., friendship ties, habits, existence or non-existence of communication channels, etc. The value of game-theoretical analysis ought not to be underestimated because of its inability to determine the most "rational" coalition structure. On the contrary, the value of the analysis is in the way it enables us to "factor out," as it were, the considerations related to the payoffs themselves and to the bargaining potentials of the players with reference to the payoffs alone; so that the *residual* factors, not covered by this analysis, but nevertheless of possibly great importance in actual situations, stand exposed.

Another example may serve to illustrate this role of game-theoretical analysis.

Michael Maschler invited a Minnesota businessman to analyze the following Three-person game:

$$v(\bar{i}) = 0 \quad (i = 1, 2, 3) \tag{6.16}$$

$$v(\overline{12}) = v(\overline{13}) = 100 \tag{6.17}$$

$$v(\overline{23}) = 50 \tag{6.18}$$

$$v(\overline{123}) = 0. \tag{6.19}$$

Application of the bargaining set principle yields the following possible configurations:

$$(75, 25, 0; \overline{12}, \bar{3}); (75, 0, 25; \overline{13}, \bar{2}); (0, 25, 25; \bar{1}, \overline{23}). \tag{6.20}$$

Maschler suggested that the businessman in the role of player 1 could keep the share of his partner (either player 2 or player 3) in a coalition down to 25 units without endangering the coalition (if the bargaining set principle of coalition stability is accepted). The businessman, according to Maschler, refused to accept this conclusion.

"I am in business all my life," he protested, "and I am telling you that you can't do business this way. You should give a guy a break." He implied that the partner, whose bargaining position was weak, should be given more than the minimum required to keep him in the coalition.

It is sometimes argued that these "extra-strategic" considerations are irrelevant to the theory. If the businessman wants to give his partner more than the theory prescribes, then, presumably, he has in mind another consideration which is not included in the situation depicted. For example, he may want to keep his partner happy or under obligation to him to insure his loyalty on *other* occasions. Or he may wish to avoid getting a repuation of being a ruthless exploiter; or he may have ethical convictions which prevent him from taking advantage of others. Such considerations, it is argued, should either be considered separately or, if they are to be included in the situation depicted by the game, appropriate modifications should be made in the payoffs. Thus, if the businessman feels bad about taking 75 units of the 100, then a negative utility should be added to the 75 to reflect his dissatisfaction with the outcome. This makes a different game with possibly a different bargaining set.

Similar arguments apply to the coalition formation. For instance, it is argued that, on the basis of given payoffs, it is impossible to say which coalition will form; and if we observe that some coalitions do tend to form rather than others, we should change the payoffs of the players accordingly. In my opinion, these arguments miss the point of game-theoretical analysis. The primary function of such analysis is not to predict, nor even to prescribe rational behavior, but to clarify and illuminate the issues in a decision problem. For example, according to the bargaining set (6.20), player 3 is entitled to the same amount, 25, whether he is in coalition with player 1 or with player 2. The bargaining set model therefore predicts no preference of partner to player 3, nor to player 2 for the same reason. Suppose, however, that we observe in experiments that coalition $(\overline{23})$ forms with greater than expected frequency. We must attribute this bias to factors not considered in the game-theoretical model; for example, to a reluctance of a player to accept substantially less than his coalition partner is getting. Or we may observe, on the contrary, that coalitions $(\overline{12})$ and $(\overline{13})$ form most frequently but that the weaker partner gets more than 25. We might conjecture that the stronger player is "making up" for the perceived inequality by yielding a side payment, i.e., ignoring a justifiable objection against his partner. In this way game-theoretic analysis provides us with a theoretic leverage for generating non-game-theoretic hypotheses.

7. *The Kernel*

The idea of the kernel of an N-person game represents another attempt to restrict the class of outcomes somehow "defensible" as characterizing rational players. We have already defined an outcome of the game as

$$(\vec{x}, \mathcal{B}) \equiv (x_1, x_2, \ldots, x_n; B_1, B_2, \ldots, B_m) \qquad (7.1)$$

in which the B's represent a partition of the N players into disjoint non-empty subsets, each encompassing the players of a coalition, while the x's represent the payoffs accruing to each of the players. We have also seen that in the context of utilities, we can suppose that $v(\bar{i}) = 0$ for each one person coalition $(\bar{i})$ and $v(B) \geqslant 0$ for each coalition B_i.

We shall assume in accordance with the principle of individual rationality that

$$x_i \geqslant 0, \quad i = (1, 2, \ldots, n) \qquad (7.2)$$

and

$$\sum_{i \in B_j} x_i = v(B_j), \qquad (7.3)$$

since each coalition can get at least its value (designated by the characteristic function) by properly coordinating the strategies of its members.

Suppose now some subset of players D contemplate forming a different coalition. That is, the members of D belong to different coalitions B_i and are considering the possibility of leaving their respective coalitions to form D. It is natural for the prospective members of the new coalition D to compare the amount they can get thereby with what they are presently getting as members of the existing coalitions B_i. In other words, they will compare v(D) with $\sum_{i \in D} x_i$. Call the difference

$$e(D) = v(D) - \sum_{i \in D} x_i \qquad (7.4)$$

the *excess* of D with respect to $(\vec{x}, \mathcal{B})$. Clearly, the larger this excess, the larger is the inducement to the members of the subset of D to form the coalition D. Note that since the members of each B_i already get jointly $v(B_i)$, it follows that

$$e(B_i) = 0, \quad i = (1, 2, \ldots, m). \qquad (7.5)$$

Now consider two players k and ℓ, both belonging to one of the coalitions B_i. Of all the *possible* coalitions (subsets) that can form, some will include both k and ℓ, and some will include only k, some only ℓ, and some neither players. Consider only those coalitions which will include k but not ℓ. This is a set of sets. We shall denote it by

$$\mathfrak{I}_{k,\ell} = \{D: D \subset N, k \in D, \ell \notin D\}. \qquad (7.6)$$

Now each of these D's will have a surplus with respect to $(\vec{x}, \mathcal{B})$ as defined by (7.4). Some of these will be positive, some negative. Since there is only a finite number of D's, some surplus (possibly several) will have the largest value. Call this value the maximum surplus of k over ℓ. Formally this maximum surplus is denoted and defined by

$$s_{k,\ell} \equiv \underset{D}{\text{Max}}\,[e(D)]. \qquad (7.7)$$

To see the significance of this number, imagine that player k is contemplating leaving the coalition of which he is presently a member and joining some other coalition. If another coalition is to contain player ℓ also, k must look forward to some arrangement with ℓ, as to how the joint payoff accruing to the new coalition will be divided. However, if the new coalition is not to contain player ℓ, then k need not consider ℓ's interest if he is contemplating joining that coalition. Fixing his attention for the moment on just one of his present coalition partners (ℓ), k is scanning all the possible coalitions which he could enter *without* ℓ. He is interested in the question which of these coalitions confers upon him the greatest advantage. If we assume that the other members of the new coalition will be satisfied with what they are presently getting, then $s_{k,\ell}$ which represents the maximal excess among all coalitions including k but excluding ℓ, will accrue entirely to k. In a way, therefore, $s_{k,\ell}$ represents player k's "hope" of what he could gain by leaving his present coalition B_i and joining another without his erstwhile partner ℓ.

Player ℓ has a similar "hope," namely of what he could gain by leaving (the same) coalition B_i and joining another without k. This is, of course, denoted by $s_{\ell,k}$.

The symbols $s_{k,\ell}$ and $s_{\ell,k}$ represent numbers which, of course, may be positive, negative, or zero. Between any two such numbers one of these relations must hold:

$$\text{Either } s_{k,\ell} > s_{\ell,k} \text{ or } s_{k,\ell} < s_{\ell,k} \text{ or } s_{k,\ell} = s_{\ell,k}. \quad (7.8)$$

Player k is said to *outweigh* player ℓ if and only if

$$s_{k,\ell} > s_{\ell,k} \quad \text{and} \quad x_\ell > 0. \quad (7.9)$$

Accordingly, ℓ outweighs k if and only if

$$s_{\ell,k} > s_{k,\ell} \quad \text{and} \quad x_k > 0. \quad (7.10)$$

Since by condition (7.2) above, no x is negative, it follows that k outweighs ℓ (written $k \gg \ell$) if and only if

$$(s_{k,\ell} - s_{\ell,k})x_\ell > 0. \quad (7.11)$$

It may, of course, happen that neither player outweighs the other. This may happen in any of the following ways:

$$s_{\ell,k} = s_{k,\ell} \qquad (7.12)$$

$$s_{k,\ell} > s_{\ell,k} \quad \text{and} \quad x_\ell = 0 \qquad (7.13)$$

$$s_{\ell,k} > s_{k,\ell} \quad \text{and} \quad x_k = 0. \qquad (7.14)$$

If any of the conditions (7.12–7.14) holds, we write $k \approx \ell$ and say that players k and ℓ are *in equilibrium*.

A coalition B_i is said to be *balanced* if every pair of players in it is in equilibrium.

Note that for player k to outweigh player ℓ, *two* conditions must be satisfied: player k's "hope" of gaining by disassociating himself from ℓ must be larger (in the sense of maximum possible gain) than the corresponding "hope" of ℓ; and, moreover, player ℓ must have some incentive for preserving the existing coalition rather than finding himself alone. This incentive is reflected in the condition $x_\ell > 0$ (since when player ℓ is alone, he gets 0).

We are now ready to define the kernel of a game. The *kernel* is the set of all individually rational payoff configurations such that every pair of players belonging to any one of the coalitions corresponding to the configuration are in equilibrium with each other.

It turns out that the kernel of a game always exists and is a subset of the bargaining set. The kernel does not depend on the labeling of the players and remains invariant under strategic equivalence. These are, of course, necessary conditions to be satisfied by any concept around which a portion of game theory is to be developed.

Let us examine the kernel of an arbitrary Three-person game. We assume as usual that $v(i) = 0$, $(i = 1, 2, 3)$ but make no further assumptions about the characteristic function, except that

$$v(\overline{12}) \leqslant v(\overline{13}) \leqslant v(\overline{23}). \qquad (7.15)$$

These inequalities imply only that we have labeled the player accordingly.

If each player is in a coalition only with himself, then the only individually rational payoff configuration is, of course, $(0, 0, 0; \overline{1}, \overline{2}, \overline{3})$. Since no player can outweigh a player who gets a zero payoff (cf. p. 127), it follows that the configuration is stable and hence in the kernel.

If two-person coalitions form, the structure of the kernel depends on whether $v(\overline{12}) + v(\overline{13}) > v(\overline{23})$ or $v(\overline{12}) + v(\overline{13}) < v(\overline{23})$; in other words, whether $v(\overline{23})$ is "predominantly" large.

First let us define three numbers $\omega_1, \omega_2, \omega_3$ which satisfy the following equations:

$$\omega_1 + \omega_2 = v(\overline{12}) \tag{7.16}$$

$$\omega_1 + \omega_3 = v(\overline{13}) \tag{7.17}$$

$$\omega_2 + \omega_3 = v(\overline{23}). \tag{7.18}$$

The set $(\omega_1, \omega_2, \omega_3)$ is called the *quota* of the game, and if player i receives ω_i, we shall say that he receives his quota. Games which admit quotas will be discussed more fully in Chapter 15.

Solving equations (7.16–7.18), we find

$$\omega_1 = 1/2[v(\overline{12}) + v(\overline{13}) - v(\overline{23})] \tag{7.19}$$

$$\omega_2 = 1/2[v(\overline{12}) + v(\overline{23}) - v(\overline{13})] \tag{7.20}$$

$$\omega_3 = 1/2[v(\overline{13}) + v(\overline{23}) - v(\overline{12})]. \tag{7.21}$$

It turns out that if $v(12) + v(13) > v(23)$, the kernel will contain the following configurations:

$$(\omega_1, \omega_2, 0; \overline{12}, \overline{3}), (\omega_1, 0, \omega_3; \overline{13}, \overline{2}), (0, \omega_2, \omega_3; \overline{1}, \overline{23}). \tag{7.22}$$

In other words, each member of a two-person coalition will get his quota, while the third player will get only his value, i.e., 0.

If, on the other hand, $v(\overline{12}) + v(\overline{13}) \leqslant v(23)$, then the kernel will contain the following configurations:

$$(0, v(\overline{12}), 0; \overline{12}, \overline{3}), (0, 0, v(\overline{13}); \overline{13}, \overline{2}), (0, \omega_2, \omega_3; \overline{1}, \overline{23}) \tag{7.23}$$

Note that if $v(\overline{12}) + v(\overline{13}) = v(23)$, then, as we can infer from equations (7.19–7.21), $\omega_1 = 0$, $\omega_2 = v(\overline{12})$, $\omega_3 = v(\overline{13})$; therefore the two kernels given in (7.22) and (7.23) coincide.

The kernel given by (7.23) indicates that in that case player 1 has no "bargaining power" vis-à-vis player 2 when the two are in a coalition. Let us see why this is so. Observe that the configuration $(0, v(\overline{12}), 0; \overline{12}, \overline{3})$ is in the bargaining set of this game (cf. Chapter 6). Player 1 has no justifiable objection against 2. If he contemplates deserting 2 in favor of 3, he must offer something to 3. He can offer him at most $v(\overline{13})$. But if $v(\overline{12}) + v(\overline{13}) \leqslant v(23)$, player 2 can certainly offer player 3 at least $v(\overline{13})$, since $v(\overline{12}) \geqslant 0$. Of course player 2, having already received the entire payoff value of the game to $(\overline{12})$, has no objection against player 1. Hence the configuration is in the bargaining set. The situation is similar to the example cited in Chapter 6 (cf. p. 120). Note that if $(\overline{23})$ forms (which is the coalition that commands the largest payoff), each player in it gets his quota in both cases considered. Their quotas, as we can see from equation (7.20) and (7.21) are always non-negative and will, in general, be positive. We might suppose, therefore, that if the kernel singles out psychologically "stable" configurations, if $v(\overline{12}) + v(\overline{13}) \leqslant v(\overline{23})$, and if the values to the two-person coalitions are positive, coalition $(\overline{23})$ is more likely to form than the other two-person coalitions. This is because in the other two cases either player 2 or player 3 cannot expect to get more than

$v(\bar{i})$, $(i = i, 2, 3)$. Since, however, this conclusion depends on psychological assumptions, it must remain only a conjecture.

Suppose, finally, that the grand coalition $(\overline{123})$ forms. Again two cases can be distinguished according to whether $v(\overline{123}) > 3v(\overline{23})$ or $v(\overline{123}) < 3v(\overline{23})$.

Case 1. $v(\overline{123}) > 3v(\overline{23})$.

Here the value of the game to the grand coalition is "preponderantly" large. In this case it turns out that the only configuration in the kernel is the one which prescribes that each member of the grand coalition shall receive an equal share of $v(\overline{123})$.

Case 2.

(i) $v(\overline{12}) + v(\overline{13}) \geqslant v(\overline{23})$ and $2v(\overline{12}) + 2v(\overline{13}) - v(\overline{23}) \leqslant v(\overline{123}) \leqslant 3v(\overline{23})$; or (ii) $v(\overline{12}) + v(\overline{13}) < v(\overline{23})$ and $v(23) \leqslant v(\overline{123}) \leqslant 3v(\overline{23})$.

Note that if both of the inequalities (i) are satisfied, the second inequality of (ii) is automatically satisfied. Therefore, in this case, the second inequality of (ii) must always hold. Roughly speaking, $v(\overline{123})$ must be large but not too large.

To calculate the (only) configuration in the kernel in this case, first compute

$$x_1 = 1/2[v(\overline{123}) - v(\overline{23})]. \tag{7.24}$$

In other words, player 1 gets one half of what he brings into coalition $(\overline{23})$ by joining it.

Next, calculate two numbers which, in a way, represent the "strengths" of players 2 and 3 in this situation:

$$w_2 = \text{Max}\,[0, v(\overline{12}) - x_1] \tag{7.25}$$

$$w_3 = \text{Max}\,[0, v(\overline{13}) - x_1]. \tag{7.26}$$

Designate by c the joint share of players 2 and 3, i.e., $1/2[v(\overline{123}) + v(\overline{23})]$. Note that the sum of the payoffs must equal $v(\overline{123})$.

Then the payoffs of players 2 and 3 will be

$$x_2 = w_2 + 1/2(c - w_2 - w_3) \qquad (7.27)$$

$$x_3 = w_3 + 1/2(c - w_2 - w_3). \qquad (7.28)$$

Example

Let $v(\overline{12}) = 2$; $v(\overline{13}) = 3$; $v(\overline{23}) = 4$; $v(\overline{123}) = 8$.

Inequalities (i) are satisfied, and this game comes under Case 2.

Applying our procedure, we obtain

$$x_1 = 1/2(8 - 4) = 2 \qquad (7.29)$$

$$c = 1/2(8 + 4) = 6;\ w_2 = \text{Max}\,[0, (2 - 2)] = 0 \qquad (7.30)$$

$$w_3 = \text{Max}\,[0, (3 - 2)] = 1. \qquad (7.31)$$

Therefore

$$x_1 = 1/2(8 - 4) = 2 \qquad (7.32)$$

$$x_2 = 1/2(6 - 1) = 5/2 \qquad (7.33)$$

$$x_3 = 1 + 1/2(6 - 1) = 7/2. \qquad (7.34)$$

Thus the configuration in the kernel, assuming the grand coalition, is (2, 5/2, 7/2).

Let us compare this result with the Shapley value solution. Using formulas (5.7–5.9) (cf. p. 109), we obtain for this game

$$\varphi_1 = 1/3(8 - 4) + 1/6(2) + 1/6(3) = 13/6 \qquad (7.35)$$

$$\varphi_2 = 1/3(8 - 3) + 1/6(4) + 1/6(2) = 16/6 \qquad (7.36)$$

$$\varphi_3 = 1/3(8 - 2) + 1/6(3) + 1/6(4) = 19/6. \qquad (7.37)$$

We see that the Shapley value solution distributes the payoffs somewhat more evenly than the kernel. To put it in another way, the kernel gives more weight to the unequal bargaining positions of the players. We have seen that under certain conditions the kernel solution awards to one member of a two-person coalition the whole value of the coalition, as if in acknowledgment that under those conditions the other partner has no bargaining power at all. The Shapley value solution awards a positive payoff to any player who is able to "bring something into a coalition." Recall our previous remark that the Shapley value is based more on considerations of equity than on power stemming from bargaining positions. Or, if one does not wish to bring considerations of equity into game theory, one could say that the Shapley value solution is based on the players' "hopes" before they try to realize these hopes in bargaining procedures. As we have seen, other analyses (e.g., the theory of the kernel) tend to shatter some of these hopes.

Case 3. $2v(\overline{23}) - v(\overline{12}) - v(\overline{13}) \leqslant v(\overline{123}) \leqslant 2v(\overline{12}) + 2v(\overline{13}) - v(\overline{23})$.

In this case, the only configuration in the kernel is

$$\left(\omega_1 + \frac{[v(\overline{123}) - \omega]}{3}, \omega_2 + \frac{[v(\overline{123}) - \omega]}{3}, \omega_3 + \frac{[v(\overline{123}) - \omega]}{3}\right) \tag{7.38}$$

where $(\omega_1, \omega_2, \omega_3)$ is the quota of the game and $\omega = \omega_1 + \omega_2 + \omega_3 = v(N)$. N.B. In the context of the generalized characteristic function ω and $v(N)$ need not be equal.

Example

We modify the game of the preceding example by letting $v(\overline{123}) = 5$, so as to satisfy inequalities defining Case 3. The values of the two-person coalitions remain the same. In this case, we readily obtain

$$\omega_1 = 1/2, \omega_2 = 3/2, \omega_3 = 5/2. \tag{7.39}$$

Thus the configuration in the kernel becomes

$$(2/3, 5/3, 8/3). \tag{7.40}$$

The Shapley value solution gives the payoff vector (7/6, 10/6, 13/6). Again we see that the Shapley value solution induces a more even distribution of payoffs than the kernel.

Case 4. $v(\overline{13}) - v(\overline{12}) \leqslant v(\overline{123}) \leqslant \text{Min}\ [v(\overline{23}), 2v(\overline{23}) - v(\overline{12}) - v(\overline{13})]$.

Here the kernel outcome is

$$(0, 1/2[v(\overline{123}) + v(\overline{12}) - v(\overline{13})], 1/2[v(\overline{123}) + v(\overline{13}) - v(\overline{12})]). \quad (7.41)$$

In other words, player 1 has completely lost his bargaining power. Note that if $v(\overline{123}) \leqslant v(\overline{23})$, player 1 has nothing to contribute by joining with players 2 and 3 already in coalition. Thereby he loses all of his bargaining power according to the kernel theory. If $(\overline{12}) > 0$ or $(\overline{13}) > 0$, player 1 would still be awarded a positive payoff by the Shapley value. If, however, $v(\overline{123}) \leqslant v(\overline{23})$, then the characteristic function is no longer super-additive and the Shapley value, which is computed only for super-additive characteristic function, no longer applies.

Case 5. $v(\overline{123}) \leqslant v(\overline{13}) - v(\overline{12})$.

The kernel outcome is $(0, 0, v(\overline{123}))$; in other words player 3 has complete control of the situation.

Example

As before, let $v(\overline{12}) = 2$, $v(\overline{13}) = 3$, $v(\overline{23}) = 4$. But now to satisfy the inequality, let $v(\overline{123}) = 1$. In this case the grand coalition will not form if a two-person coalition forms, since the two-person coalition can only lose by admitting another member. Player 3 holds all the cards, since he can always wean either 1 or 2 away from the coalition $(\overline{12})$.

It is interesting to observe that when $v(N)$ is sufficiently large—specifically, three times the largest of the three values $v(\overline{ij})$—the only disbursement in the kernel (assuming the grand coalition) awards an equal payoff to all three players. In these situations the kernel solution is even more "egalitarian" than the Shapley value. Conversely, when v(N) becomes sufficiently small, first one, then two of the players are totally deprived of positive payoffs in the grand coalition. Roughly, when inducement to cooperation is large, an egalitarian disbursement is in some sense stable; when the inducement for cooperation among the three actually becomes "negative," at first one, then two players lose their bargaining power. All power is concentrated in the "strongest" player.

We have seen (cf. p. 89) that a Three-person game has a core if $v(N) \geqslant 1/2[v(\overline{12}) + v(\overline{13}) + v(\overline{23})]$. That is, if this inequality holds, there exist imputations which will make the players "happy" in the sense that they will not want to leave the grand coalition. The condition which insures an egalitarian imputation in the kernel (cf. Case 1) is even stronger, since if $v(\overline{23})$ is the largest of the $v(\overline{ij})$, we must have $3v(\overline{23}) > [v(\overline{12}) + v(\overline{13}) + v(\overline{23})] > 1/2[v(\overline{12}) + v(\overline{13}) + v(\overline{23})]$, hence certainly $v(N) \geqslant 1/2[v(\overline{12}) + v(\overline{13}) + v(\overline{23})]$, since $v(N) > 3v(\overline{23})$.

Returning to the result which assigns v(N) to the strongest player in the kernel, it certainly violates our intuitive notion of rationality. If v(N) is actually smaller than some $v(\overline{ij})$ [in our example it was assumed smaller than every $v(\overline{ij})$], why should the grand coalition form at all? We should keep in mind, however, that the configurations in the kernel are computed for each *given* coalition structure. It singles out the stable payoff disbursement for *that* structure. Whether that structure occurs is another question. One way of looking at it is to imagine that the given coalition structure is *prescribed* to the players unless at least one can prove that it is not "fair." In the context of kernel theory, this would mean that the player *outweighs* another player in the same coalition. Only then is the coalition allowed to dissolve. Now, according to the definition given above, no player can be outweighed if his payoff is zero. Thus player 3 cannot outweigh either of the other players with respect to the configuration $[0, 0, v(N)]$. It turns out also that neither of them can outweigh him. Hence the three players are in equilibrium and must accept the result.

A similar way of looking at these situations is to imagine that if no agreement results, each player must play by himself and, as a consequence, will get $v(\bar{i}) = 0$. Such outcomes are certainly conceivable even if several much better options are open to the players. This being the case, the outcome $[0, 0, v(N)]$, where $v(N) \leqslant v(ij)$, although not "rational," is certainly conceivable: neither player 1 nor player 2 can gain anything by dissolving the coalition.

We have already seen how the results of strict strategic analysis and conjectures derived from other intuited notions are often at variance. As long as we are not thinking in the context of a normative or a prescriptive behavioral theory, i.e., as long as we confine ourselves strictly to pursuing the consequences of explicit formal assumptions, these discrepancies do not matter. Once we pass into the realm of a descriptive behavioral theory, then the question whether strategic analysis or psychological factors should be given a greater weight in constructing models of behavior can be answered only on empirical grounds. Only comparisons of models with experiments or observations can decide it. If, on the other hand, we wish to construct a *normative* theory, i.e., be in position of advising "rational players" on what the outcomes of a game "ought" to be, we see that we cannot do this without further assumptions about what explicitly we mean by "rationality." Intuitive notions or dictionary type definitions are of no help here.[15]

8. Restrictions on Realignments

As has been pointed out, N-person game theory in characteristic function form abstracts not only from the actual rules of the game (embodied in the extensive form) but also from the strategies (embodied in the normal form) available to the players. What is left in the characteristic function is only information about what each of the $2^N - 1$ non-empty subsets of players can be sure to get if they played as a coalition vis-à-vis the remaining players (in a counter-coalition).

If players are faced with a game given only in its characteristic function form, they are not concerned with *how* they can get what the characteristic function gives them individually and collectively in playing the original game. In this context, our remarks to the effect that the concept of imputation implies a grand coalition, or at least a coalition which can get v(N), become irrelevant. When the connection with the original game is severed, no contradiction is involved in assuming that the players jointly get v(N), even though they are partitioned into k coalitions in such manner that, were the game played as a k-person non-cooperative game, the players could not jointly get v(N).

We now define the N-person game as follows. A certain amount is to be divided among the players. Next, it is specified how much each subset of players could get if they simply declared themselves to be a coalition, regardless of what others get or how they are organized. The problem to be solved is what to expect in the way of coalition formation and apportionment of payoffs.

Note that the idea of the core is a reasonable solution of this problem if the characteristic function is such that the core is not empty. The core solution says that the amount v(N) will be apportioned in such a way that no subset will be motivated to leave the grand coalition by what the characteristic function awards to them as a coalition of their own.

We have seen also that the weakness of this concept of solution is the circumstance that many games have empty cores, including all constant-sum N-person games. When the core is empty, it means that whatever coalition structure obtains, there is always some subset of players, not all in the same coalition, who would rather be in a coalition, since they can get more that way than with the existing structure. If no restrictions are put on the way coalitions are formed, the disgruntled subset of players *can* form a coalition and so disrupt the old structure. But in the new structure there is certain to be another subset of players, not all in the same coalition, who would rather be in a coalition for the same reason as before. In other words, no coalition structure is stable if the core is empty.

A way to get around this difficulty is to impose restrictions on the way *changes* in coalition structures are made. Assume that v(N) is to be partitioned among the n players and that temporarily the players have coalesced into a number of coalitions (B_1, B_2, . . . , B_k). We ignore for the moment how the coalition structure came about. We are now concerned only with how it can change. In order to increase the stability of coalition structures, once they are formed, we shall introduce some restrictions on the possible ways in which the coalition structure can change. An example of such restriction could be the following: the only way the players can realign themselves is by the process of either expelling or admitting no more than one member from or to one coalition. Or, to specify a somewhat more liberal rule: the only realignments allowed are those which occur when a member is admitted or a member expelled or both simultaneously.

The problem to be solved now is the following. Given a coalition structure and the rule governing realignments, what disbursements of payoffs form a *stable pair* with a given coalition structure, in the sense that there is no inducement for any of the players to realign themselves into any other coalition structure *allowed* by the rules of realignment?

Let $\mathfrak{I}$ stand for any of the possible *partitions* of n players into coalitions. Thus the domain of $\mathfrak{I}$ is the set of all possible partitions of n players. Let $\psi(\mathfrak{I})$ be a function on partitions to sets of partitions. Thus, given a partition $\mathfrak{I}$, the value of $\psi(\mathfrak{I})$ is some set of partitions. The idea is that these are the partitions which are allowed as realignments of players originally partitioned by $\mathfrak{I}$. We assume that $\mathfrak{I}$ is always a member of the set determined by $\psi(\mathfrak{I})$, i.e., that "no realignment" is always allowed.

Let now $(x_1, x_2, \ldots, x_n)$ be an imputation of the N-person game, i.e., a vector satisfying

$$x_i \geqslant v(\bar{i}) \tag{8.1}$$

$$\sum_{i=1}^{n} x_i = v(N). \tag{8.2}$$

We shall impose a minor restriction on all $\mathfrak{I}$; namely if $x_i = v(\bar{i})$, we shall suppose that player i is in a coalition only with himself. The restriction is a consequence of assuming that it takes *some* inducement to get a player to join any coalition with someone else. The pair $(\vec{x}, \mathfrak{I})$, i.e., the pair consisting of the imputation and the partition, will be called ψ-stable pair if and only if

$$\sum_{i \in S} x_i \geqslant v(S) \tag{8.3}$$

for all S which are coalitions in the set of partitions $\psi(\mathfrak{I})$.

That is to say, once the coalition structure has formed in accordance with some partition $\mathfrak{J}$, it forms a stable pair with some imputation if there is no inducement for any subset S of players who *could* form a *permitted* coalition [expressed in the fact that S is a coalition in one or more of the partitions of $\psi(\mathfrak{J})$] to actually form such a coalition.

Nothing is said in this formulation concerning how the coalition structure itself came into being. We might suppose that it came into being by initial contacts among the players. The question is now whether, once the coalition structure has formed, the imputation $\vec{x}$ is acceptable to the players so partitioned. Roughly speaking, it is so if no subset of players can do better under the restrictions imposed by $\psi(\mathfrak{J})$. Note that if $\psi(\mathfrak{J})$ comprises *all* possible partitions, then only the imputations of the core will be acceptable. Indeed, these imputations always form stable pairs with any partition.

Further, nothing is said about what will happen if $(\vec{x}, \mathfrak{J})$ is not a stable pair. Conceivably, some other coalition structure $\mathfrak{J}'$ might form according to the changes allowed under the rule $\psi(\mathfrak{J})$. Among the new coalitions, there will presumably be included a coalition S in $\psi(\mathfrak{J})$, motivated by the fact that the members of S can demand more than they can jointly get in the imputation $\vec{x}$. Suppose this happens, and the players are presented with another imputation $\vec{x}'$, in which the demands of the members of S were satisfied. A new pair $(\vec{x}', \mathfrak{J}')$ results, and the question of *its* stability is raised again. This process might continue until a stable pair $(\vec{x}^*, \mathfrak{J}^*)$ is reached. Then there will no longer be any inducement for the coalition structure to change. The theory of ψ-stability says nothing about how the "search" for a stable pair actually goes on. The model (like practically all the models used in game theory) is a static one, not a dynamic one. It singles out "equilibrium states" of a system without specifying how such states are reached.

As an illustration, consider a Four-person constant-sum game with characteristic function

$$v(\bar{i}) = 0 \quad (i = 1, 2, 3, 4); \qquad v(\overline{12}) = 3;\ v(\overline{13}) = 2;\ v(\overline{14}) = 1 \tag{8.4}$$

$$v(\overline{34}) = 1;\ v(\overline{24}) = 2;\ v(\overline{23}) = 3; \qquad v(N - \{i\}) = 4 \quad (i = 1, 2, 3, 4) \tag{8.5}$$

$$v(N) = 4. \tag{8.6}$$

The possible coalition structures are

$$\begin{array}{llll} \mathfrak{J}_1: & \{1, 2, 3, 4\} & \mathfrak{J}_9: & \{\overline{13}, \overline{24}\} \\ \mathfrak{J}_2: & \{\overline{12}, 3, 4\} & \mathfrak{J}_{10}: & \{\overline{14}, \overline{23}\} \\ \mathfrak{J}_3: & \{\overline{13}, 2, 4\} & \mathfrak{J}_{11}: & \{\overline{234}, 1\} \\ \mathfrak{J}_4: & \{\overline{14}, 2, 3\} & \mathfrak{J}_{12}: & \{\overline{134}, 2\} \\ \mathfrak{J}_5: & \{1, \overline{23}, 4\} & \mathfrak{J}_{13}: & \{\overline{124}, 3\} \\ \mathfrak{J}_6: & \{1, \overline{24}, 3\} & \mathfrak{J}_{14}: & \{\overline{123}, 4\} \\ \mathfrak{J}_7: & \{1, 2, \overline{34}\} & \mathfrak{J}_{15}: & \{\overline{1234}\} \\ \mathfrak{J}_8: & \{\overline{12}, \overline{34}\} & & \end{array} \tag{8.7}$$

We must now specify a rule according to which changes in the coalition structure can be made. Let us assume the following rule: (1) A coalition may expel a member; (2) A coalition may recruit a member but only provided he is not in a coalition with anyone else (i.e., "raiding" an established coalition is forbidden); (3) Changes allowed by (1) and (2) may occur only one at a time. As has been said, "no change" is always allowed.

Now given the coalition structures ($\mathfrak{J}_i$), the coalition structures allowed by the above rules are determined. For example,

$$\psi(\mathfrak{J}_1) = \{\mathfrak{J}_1, \mathfrak{J}_2, \mathfrak{J}_3, \mathfrak{J}_4, \mathfrak{J}_5, \mathfrak{J}_6, \mathfrak{J}_7\} \tag{8.8}$$

i.e., any two players can combine into a coalition.

$$\psi(\mathfrak{J}_4) = \{\mathfrak{J}_1, \mathfrak{J}_4, \mathfrak{J}_{13}, \mathfrak{J}_{12}, \mathfrak{J}_{10}\} \tag{8.9}$$

i.e., the coalition $(\overline{14})$ may break up, or it may invite either player 2 or player 3 to join it or players 2 and 3 may join in a coalition.

$$\psi(\mathfrak{J}_{11}) = \{\mathfrak{J}_7, \mathfrak{J}_6, \mathfrak{J}_5, \mathfrak{J}_{11}, \mathfrak{J}_{15}\} \tag{8.10}$$

$$\psi(\mathfrak{J}_8) = \{\mathfrak{J}_2, \mathfrak{J}_7, \mathfrak{J}_8\}. \tag{8.11}$$

In the last case, either of the two coalitions may break up.

We can now test any pair for ψ-stability as defined.

We note first that the only coalition structures which can form stable pairs with some imputations are those in which three players are in a coalition against one, and those consisting of two pairs. This is because, if there is more than one isolated player, and at least one of them gets a positive payoff, the other can be absorbed into a coalition of three, which can take the entire amount $v(N)$, thus getting more than they would in the imputation. On the other hand, if two isolated players get 0, they can combine in a coalition to get more. We are now in a position to list the stable pairs.

$\mathfrak{J}_8$ forms a stable pair with any imputation where $x_1 + x_2 \geqslant 3$ and $x_3 + x_4 \geqslant 1$. Actually these two inequalities can be satisfied simultaneously only if they are equalities, since otherwise the sum of the payoffs would exceed $v(N)$, which cannot occur in an imputation.

Similar conditions must be satisfied for all the other two-pair structures. As for three-against-one structures, clearly the lone player must get 0, while every pair of the triple must get at least their value.

For example, the imputation $(1, 2, \frac{1}{2}, \frac{1}{2})$ is the only imputation which forms a stable pair with coalition structure $\mathfrak{I}_8 = \{(\overline{12}), (\overline{34})\}$. Several imputations form stable pairs with coalition structure $\mathfrak{I}_{11} = \{(1), (\overline{234})\}$, for example, (0, 1.7, 1.7, 0.6), (0. 1.2, 2,0, 0.8), etc. Note that when two-pair coalitions form, only one imputation forms a stable pair with the resulting coalition structure; but when the split is three-against-one, there is a wide latitude in the imputations which form a stable pair. It is hard to say whether this suggests that three-against-one coalitions are more likely to form than two-against-two coalitions. On the one hand, one might suppose so, because many more imputations are able to make the coalition members "happy," in the sense that they can do no better in another coalition. On the other hand, one might argue that there is more opportunity for disputes about the final disbursements of payoffs in the three-against-one situation (since many different imputations are consistent with it, while only one imputation is consistent with the two-against-two structure).

The Shapley value of this game turns out to be (1, 4/3, 1, 2/3). It awards at least $v(\overline{ij})$ to the pair-coalitions $(\overline{13})$, $(\overline{14})$, $(\overline{24})$, and $(\overline{34})$, but not to $(\overline{12})$ or $(\overline{23})$. Therefore, it will form a stable pair only with $\mathfrak{I}_9$. It will not form a stable pair with any three-against-one partition, because any three players can get jointly $v(N)$, which the Shapley value imputation does not give them.

It must be kept in mind that all the results we have obtained are with respect to the particular function $\psi(\mathfrak{I})$ chosen; that is, under particular restrictions on the realignment of the players in coalition. Recall that if there are no restrictions on realignment, *no* stable pairs exist in a constant-sum game. For in this case the imputations which form stable pairs must be in the core, and we have seen that constant-sum games have empty cores.

The restrictions on realignments may be taken to represent something like "social friction." It is well known that no matter how "free" competition may be for customers or for goods and services, economic conditions can never be in perfect equilibrium. People cannot travel arbitrary distances to buy things where they are cheaper. In the language of game theory, they are not able to form coalitions with sellers who are too far away. It takes effort to recruit people into associations, unions, political movements, etc., even if it is in their interests to join them. Traditions, sentiments, historical inertia keep alliances from forming or from dissolving regardless of the immediate interests of the actors. All these are examples of "social friction," conditions which preclude the development of an applicable theory in vacuo. The theory of ψ-stability is an attempt to capture some presumably essential features of these phenomena and to formalize them as rigorous assumptions super-imposed on the "classical" theory of the N-person game.[16]

We shall return to this model in the context of a small (simple) market (Chapter 12).

9. *Games in Partition Function Form*

The basic idea leading to the characteristic function form of an N-person game rests on the assumption that if some k of the n players ($k < n$) form a coalition K, the remaining $n - k$ players will form a counter-coalition $N - K$. This assumption is justified by an implicit assumption of group rationality. The remaining players have nothing to lose and possibly something to gain by forming a counter-coalition, for they can *jointly* get at least as much and possibly more by doing so as by not doing so. Implicit in this justification is also another assumption which is, perhaps, already implied by the clause "If k players form a coalition . . . ," namely that this coalition has "jelled," as it were, and is not subject to disruption. Otherwise there may be better things for the remaining $n - k$ players to do than form a counter-coalition; for example, they (or some subset of them) may try to disrupt the coalition K by luring members away. They may well be in a position to do so if they can promise some members of K ultimate payoffs (i.e., components in an imputation) which are larger than those the latter can expect to get as members of K. A counter-coalition may also fail to form because it involves excessive costs (of establishing or maintaining communications); or it may put its members under social opprobrium (unethical collusions); or it may be illegal (cartels, trusts).

The extremely "fluid" nature of reasoning about games in characteristic function form stems, it seems, from this amalgam of assumptions. On the one hand, the value assigned to a coalition is derived from the assumption that its members have merged into a single player, playing against his counterpart represented by the complement coalition. On the other hand, in comparing imputations, the situation is imagined as allowing any subset of players to consider themselves as a potential coalition. While there is no actual contradiction here (since *potential coalitions* may well be factors in the bargaining among the members of the grand coalition), the idea fails to capture the influence on the outcome of the game by the coalitions which *actually* form.

We have already examined some modifications of the classical theory, in particular those where restrictions are in effect placed on the way the players can shift their loyalties. In the approach construed in the ψ-stability concept, for example, the changes in the coalition structure are restricted; in the approach based on the bargaining set and the kernel, a coalition structure once formed is assumed fixed while the players do their bargaining. That is, before a coalition structure and an associated disbursement of payoffs are declared to be "unstable," both the objections and the counter-objections of players against each other must be "heard."

These treatments depart from the Von Neumann-Morgenstern theory in two fundamental ways. First, coalition structures are considered which are not simply dichotomies (two complementary coalitions). Second, in the case of the theories derived from the bargaining set and the kernel, the disbursements of payoffs are not confined to imputations. In fact, in some of these theories, even the super-additive property of the characteristic function is dropped.

The extension of possible payoff disbursements beyond the set of imputations seems to be a logical consequence of the very structure of non-constant-sum games. By definition, the total sum of the payoffs in an imputation $\left(\sum_{i=1}^{n} x_i\right)$ is what the players could get if they played as a grand coalition. In the case of a constant-sum game, this sum represents also what the players would jointly get if a coalition played against a complementary coalition, in which case each coalition would get its full value determined by the characteristic function. Therefore, in this context, it seems relevant to ask how the players would divide up the joint payoff of an imputation. Suppose, however, the players of a non-constant-sum game are partitioned into coalitions. In general, they will *not* get jointly the amount guaranteed by an imputation. Moreover, if coalition structures other than dichotomies (i.e., consisting of more than two coalitions) occur, a coalition may well receive jointly more than the value of the game assigned to it by the characteristic function. This value, recall, is assigned to a coalition on the supposition that the remaining players form a complementary coalition which attempts to hold the payoff of the first coalition to a minimum. In the case of a constant-sum game, this amount is just what the coalition can expect. (It can expect to get more if the remaining players fail to unite.) In non-constant-sum games it may conceivably get more, even playing against a complementary coalition, since, if the complementary coalition tries to get as much as it can, it can in the process also give more to the first coalition (cf. p. 82).

It makes sense, therefore, to assign to each subset of S players not a single joint payoff, as is done in the characteristic function form of the game, but a *set* of joint payoffs, one for each coalition structure, i.e., each set of sets which includes S as a member. This assignment characterizes the *partition function representation* of the game.

The *value* assigned by the partition function to each coalition is the smallest of this set of joint payoffs. Thus, if the game is a constant-sum game, the value of the game to a coalition will coincide with the value assigned to it by the characteristic function derived from the game in question. In a constant-sum game this is the outcome of the worst thing that can happen to a coalition, namely the formation of a counter-coalition. In non-constant-sum games, however, this may not be the case. It may happen that the counter-coalition in pursuit of their collective interest will reach an outcome which will give the first coalition more than if the remaining members pursued their self-interest individually.

As in the case of games in characteristic function form, it is customary to abstract from the game itself; so that the values can be assigned to each coalition arbitrarily.

Consider a Four-person game. There are 15 possible partitions (see p. 141). Consider the coalition $(\overline{12})$. In the actual game the joint payoff which they will get will depend on whether players 3 and 4 will combine into a coalition. Similarly, player 1 will get different amounts depending on whether players 2, 3, and 4 play separately, or whether 2 and 3 form a coalition, but not 3 and 4, etc. A different possible payoff is assigned to each player in each case. The smallest of these is the *value* of the game to player 1.

Example

Consider the Three-person Prisoner's Dilemma game discussed in Chapter 2. The set of all partitions (as in any Three-person game) is

$$\begin{aligned} P_0 &= (\overline{123}) \\ P_1 &= (\overline{1}, \overline{23}) \\ P_2 &= (\overline{2}, \overline{13}) \\ P_3 &= (\overline{3}, \overline{12}) \\ P_4 &= (\overline{1}, \overline{2}, \overline{3}). \end{aligned} \tag{9.1}$$

Our analysis of the game in extensive form shows that we can assign the following outcome functions:

$$F_{P_0}(\overline{123}) = 3 \quad (9.2)$$

$$F_{P_i}(\bar{i}) = 2,\ F_{P_i}(\overline{jk}) = 0 \quad (i, j, k = 1, 2, 3, \text{ all distinct}) \quad (9.3)$$

$$F_{P_4}(\bar{i}) = -1 \quad (i = 1, 2, 3). \quad (9.4)$$

Then, by our definition of value,

$$v(\bar{i}) = -1 \quad (i = 1, 2, 3) \quad (9.5)$$

$$v(\overline{ij}) = 0 \quad (i, j = 1, 2, 3; i \neq j) \quad (9.6)$$

$$v(\overline{123}) = 3. \quad (9.7)$$

In this example, the function $v(\ )$ is super-additive. It need not be super-additive, however. In the context of games in partition function form (as in the context of the bargaining set and the kernel), we may have $v(S) + v(T) > v(S \cup T)$ for $S \cap T = \emptyset$.

As in the case of the Von Neumann-Morgenstern theory, there is no loss of generality in assuming that $v(\bar{i}) = 0$ $(i = 1, 2, \ldots, n)$.[17] An imputation will now be defined as a payoff vector $\vec{x} \equiv (x_1, x_2, \ldots, x_n)$ if and only if

$$x_i \geqslant v(\bar{i}) = 0 \quad (9.8)$$

$$\sum_{i=1}^{n} x_i = \sum_{P_j \in P} F_P(P_j) \text{ for } some \text{ P}. \quad (9.9)$$

The first of these conditions corresponds to the previous definition of imputation. The second is a generalization. It requires that the disbursement of payoffs to which the status of imputation is accorded be realizable for *some* coalition structure. In the previous contexts, only the disbursements of payoffs whose sum was equal to $v(N)$, i.e., the value of the grand coalition, were called imputations. In the present theory, any disbursement is an imputation if the sum of the payoffs to all coalitions, assigned by the partition function in some coalition structure, equals the sum of the disbursed payoffs.

In our Three-person Prisoner's Dilemma game (2, 0, 0) is now an imputation even though $\sum_{i=1}^{n} x_i = 2 < 3 = v(\overline{123})$, because equation (9.9) is satisfied for P_1, P_2, and P_3. Similarly $(-1, -1, -1)$ is an imputation, because equation (9.9) is satisfied for P_4.

We now define domination in the context of games in partition function form. Recall that in the context of the classical theory an imputation $\vec{x}$ was said to dominate imputation $\vec{y}$ *via* a subset S of players (written $\vec{x} \underset{S}{\text{dom}} \vec{y}$) if

$$x_i > y_i \text{ for all } i \in S \tag{9.10}$$

$$\sum_{i \in S} x_i \leqslant v(S). \tag{9.11}$$

In the present context, $\vec{x} \underset{S}{\text{dom}} \vec{y}$ if inequalities (9.10) and (9.11) are satisfied, and *in addition* if

$$\sum_{i \in S} x_i = \sum_{P_j \in P} F_P(P_j) \text{ for some } P \text{ with } S \in P. \tag{9.12}$$

The additional condition is the *realizability* of $\vec{x}$. To be sure, $\vec{x}$ is an imputation; but it is necessary for $\vec{x}$ to be among the imputations arising from a partition in which S is actually a coalition.

In our Prisoner's Dilemma game, $\vec{x} = (1, 1, 1)$ and $\vec{y} = (-1, -1, -1)$ are imputations. $\vec{x}$ dominates $\vec{y}$ via $(\overline{123})$. But $\vec{x}$ does not dominate $\vec{y}$ via $(\overline{12})$, because $\vec{x}$ is not among the imputations when 1 and 2 are in a coalition. On the other hand, $\vec{z} = (0, 0, 2)$ does dominate $\vec{x}$ via $(\overline{12})$, because $\vec{z}$ is among the imputations when 1 and 2 are in coalition.

As in the classical theory, $\vec{x}$ is said to (simply) dominate $\vec{y}$ if for some subset S of N, $\vec{x}$ dominates $\vec{y}$ via S.

The definition of solution and core are exactly the same as in the classical theory. A set of imputations K is a solution if no imputation in K dominates any other imputation in K, and if for every imputation $\vec{z}$ not in K there is at least one imputation in K which dominates $\vec{z}$. The core is the set of imputations not dominated by any imputation.

Let us find the solutions of the Three-person Prisoner's Dilemma. We shall normalize the game, so that $v(\bar{i}) = 0$ $(i = 1, 2, 3)$ and $v(\overline{123}) = 1$. Then equations (9.2)–(9.4) become

$$F_{P_0}(\overline{123}) = 1 \tag{9.13}$$

$$F_{P_i}(\bar{i}) = 1/2;\ F_{P_i}(\overline{jk}) = 1/3 \quad (i, j, k = 1, 2, 3). \tag{9.14}$$

$$F_{P_4}(\bar{i}) = 0 \quad (i = 1, 2, 3). \tag{9.15}$$

There are three sets of imputations:

(1) The set C_0, in which $\sum_{i=1}^{3} x_i = 1$ (realizable only if the grand coalition has formed).

(2) The set C_1, in which $\sum_{i=1}^{3} x_i = 5/6$ (realizable only if a pair of players has formed a coalition.

(3) The set G, in which $\sum_{i=1}^{3} x_i = 0$ (realizable only if each player plays alone).

Of course, we must also have $x_i \geqslant 0$ in all cases, since $v(\bar{i}) = 0 (i = 1, 2, 3)$.

The set G has only one imputation. It does not dominate any other; so it cannot be in a solution.

Now we shall show that the imputations in C_1 cannot be in a solution. Consider the imputation $x_0 = (1/2, 1/4, 1/4)$, which is in C_0. No imputation in C_1 can dominate it via a single player, because a single player in a coalition with himself can get no more than 0; and so inequality (9.10) cannot be satisfied, since every player gets a positive payoff in x_0. Nor can a pair of players, acting in a coalition, get more than 1/3, whereas they get at least 1/2 in x_0. Therefore no imputation in C_1 can dominate x_0 via a pair of players. Finally, all three players cannot prefer any imputation in C_1 to x_0, because all three get jointly more in x_0. On the other hand, there exists no imputation in C_0 which dominates any other imputation in the same set. Also an imputation in C_0 dominates all imputations not in C_0 via $(\overline{123})$. Therefore C_0 qualifies as a solution, and it is easy to see that there are no others.

The unique solution of this game can be represented geometrically. It is the entire simplex defined by $\sum_{i=1}^{3} x_i = 1$, $x_i \geqslant 0$ $(i = 1, 2, 3)$; in other words, the triangle which is the portion of plane $x_k + x_2 + x_3 = 1$ in the first $(+++)$ octant of the three-dimensional space R^3.

All Three-person games in partition function form can be classified into four categories or genera, depending on the magnitudes of the parameters of their partition functions. For the general Three-person game, we have, in fact,

$$F_{P_0}(\overline{123}) = c \tag{9.16}$$

$$F_{P_i} = d_i;\ F_{P_i}(\overline{jk}) = e_i \quad (i, j, k = 1, 2, 3, \text{ all distinct}) \tag{9.17}$$

$$F_{P_4}(i) = g_i \quad (i = 1, 2, 3). \tag{9.18}$$

Assuming as usual $v(\overline{i}) = 0$, we see that, in view of the definition of value in this context, either d_i or g_i

must vanish. Whichever vanishes, the other must be non-negative. Denote $d_i + e_i$ by c_i. Then the imputation vectors fall into five classes:

$$C_0:\quad \text{all } (x_1, x_2, x_3) \text{ in which } \sum_{i=1}^{3} x_i = c. \tag{9.19}$$

$$C_1:\quad \text{all } (x_1, x_2, x_3) \text{ in which } \sum_{i=1}^{3} x_i = c_1. \tag{9.20}$$

$$C_2:\quad \text{all } (x_1, x_2, x_3) \text{ in which } \sum_{i=1}^{3} x_i = c_2. \tag{9.21}$$

$$C_3:\quad \text{all } (x_1, x_2, x_3) \text{ in which } \sum_{i=1}^{3} x_i = c_3. \tag{9.22}$$

$$C_4:\quad \text{all } (x_1, x_2, x_3) \text{ in which } \sum_{i=1}^{3} x_i = \sum_{i=1}^{3} g_i. \tag{9.23}$$

The games, then, fall into the following genera (we label the players so that $c_1 \geqslant c_2 \geqslant c_3$):

Genus 0. In these games $c > c_1 \geqslant c_2 \geqslant c_3$. That is, the parameter associated with the imputations arising in the grand coalition is larger than any of the parameters associated with the imputations arising when there are pair coalitions.

The games of this genus all have a single solution, namely the set of imputations in C_0. Our Three-person Prisoner's Dilemma was a game of this sort. They coincide with the entire simplex $\sum_{i=1}^{3} x_i = c,\ x_i \geqslant 0$. We shall denote by $A(b)$ the simplex defined by $\sum_{i=1}^{3} x_i = b,\ x_i \geqslant 0$. Thus the (unique) solution of the games of genus G_0 is the simplex $A(c)$.

Genus 1. In these games $c_1 \geqslant x > x_2 \geqslant c_3$. The (unique) solutions of these games are the sets of points

$$A(c_1) - \{\vec{x} \mid x_2 + x_3 < e_1\}, \tag{9.24}$$

that is, the points of $A(c_1)$ *except* those for which $x_2 + x_3 < e_1$. This solution is represented by the shaded area in Figure 12.

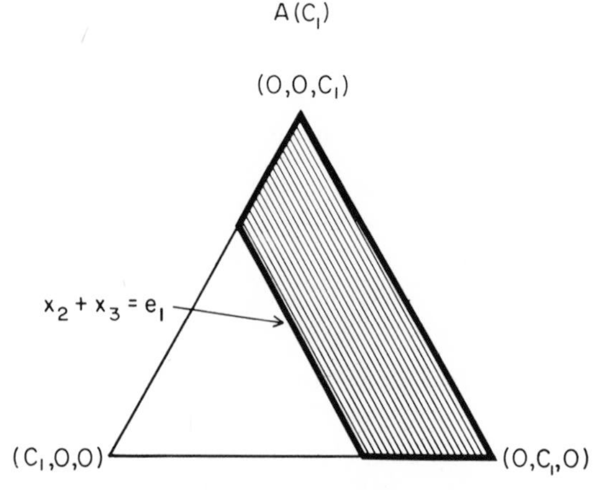

FIG. 12. The (unique) solution of games of Genus 1 is the set of imputations within and on the boundary of the shaded area.

Genus 2. This genus consists of two "species" of games, namely

$$\text{Species 2A, where } c_1 > c_2 \geqslant c > c_3 \qquad (9.25)$$

$$\text{Species 2B, where } c_1 = c_2 \geqslant c > c_3. \qquad (9.26)$$

The (unique) solutions of the games of Species 2A are unions of two sets of points, one set being on the simplex $A(c_1)$, the other on the simplex $A(c_2)$.

The points on simplex $A(c_1)$ are represented by

$$K_1\colon\ A(c_1) - \{\vec{x} \mid x_2 + x_3 < e_1\} - \{\vec{x} \mid x_2 + x_3 > e_1 + c_1 - c_2,\ x_1 + x_3 < e_2\}, \qquad (9.27)$$

i.e., those which remain after the sets represented by the braces have been deleted.

The points on simplex $A(c_2)$ are represented by

$$K_2\colon\ A(c_2) - \{\vec{x} \mid x_2 + x_3 < e_1\} - \{\vec{x} \mid e_1 < x_2 + x_3,\ x_1 + x_3 < e_2\}. \qquad (9.28)$$

In other words, the points on simplex $A(c_2)$ are those

where $x_2 + x_3 = e_1$ *and* those where $x_2 + x_3 > e_1$ but $x_1 + x_3 \geqslant e_2$.

The solution is the set

$$K \equiv K_1 \cup K_2, \qquad (9.29)$$

which is shown in Figure 13.

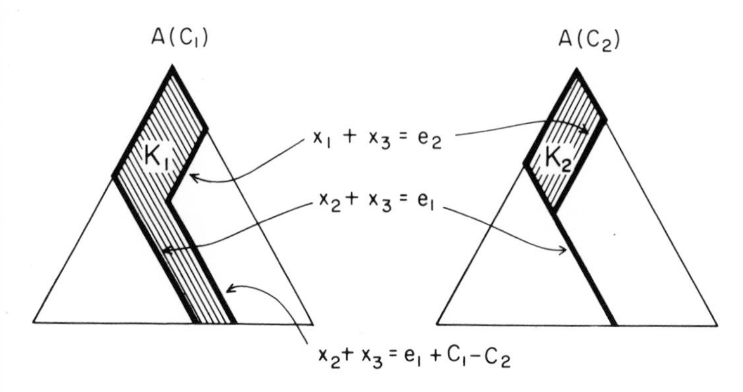

FIG. 13. The (unique) solution of games of Genus 2A is the union of the two sets (shaded areas and thick lines).

The solutions of games in Species 2B are no longer unique. They comprise the imputations contained in the parallelogram K_2 defined by

$$A(c_1) - \{\vec{x} \mid x_2 + x_3 < e_1\} - \{\vec{x} \mid x_1 + x_3 < e_2\}. \qquad (9.30)$$

Each solution, however, includes an additional set of points, defined as follows. Draw a curve K_1 on $A(c_1)$ from the point $(d_1, d_2, c_1, -d_1, -d_2)$ to the edge $x_3 = 0$, such that its coordinates x_1 and x_2 are nondecreasing as x_3 decreases. The parallelogram K_2 *and* the points on any such curve K_1 constitute a solution:

$$K = K_1 \cup K_2, \qquad (9.31)$$

as shown in Figure 14.

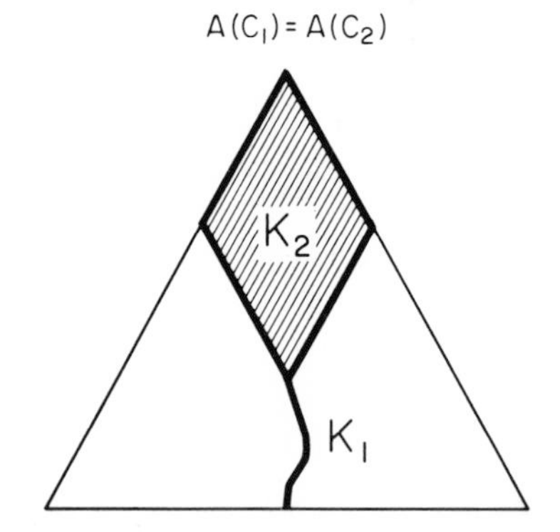

FIG. 14. Typical solution of a game of Genus 2B.

Genus 3. This genus consists of four species:

$$\text{Species 3A, where } c_1 > c_2 > c_3 \geqslant c; \qquad (9.32)$$

$$\text{Species 3B, where } c_1 = c_2 > c_3 \geqslant c; \qquad (9.33)$$

$$\text{Species 3C, where } c_1 > c_2 = c_3 \geqslant c; \qquad (9.34)$$

$$\text{Species 3D, where } c_1 = c_2 = c_3 \geqslant c. \qquad (9.35)$$

The solutions of these games are rather involved. That is, they are represented geometrically by rather irregular configurations and set-theoretically by complex formulas. The interested reader is referred to Thrall's original work.[18]

An important result of partition function theory was proved by W. F. Lucas.[19] It concerns the solution of a Four-person game with distinct imputation simplexes; i.e., those in which each partition of the players (coalition structure) determines a simplex of imputations which is not identical with any of the simplexes determined by other partitions.

If this condition is satisfied, Lucas' theorem asserts that the "solution of the game is unique and is polyhedral in nature." The uniqueness of the solution does not, of course, single out a unique imputation. Rather it singles out a set of imputations. This set comprises the points on the faces and inside a polyhedron in the payoff space, hence a restricted portion of that space.

Note that all of the Three-person games of Genus 0 and Genus 1 have unique "polyhedral" solutions even though their imputation simplexes may not be distinct. (A triangle and a trapezoid are two-dimensional "polyhedra.") Lucas' result states that Four-person games in which all of the simplexes *are* distinct *all* have unique polyhedral solutions. (In this case, the polyhedra are, of course, "genuine" three-dimensional polyhedra; e.g., parallelepipeds, pyramids, or any convex portions of space bounded by planes.)

This result might suggest a conjecture that any N-person game with distinct imputation simplexes has a unique polyhedral solution. (The conjecture might appear plausible in view of the fact that the imposed condition of distinct imputation simplexes is pretty strong.) However, the conjecture turned out to be false. Lucas cited a counter-example of a Five-person game in which all of the simplexes were distinct but for which the conclusion was shown to be false.

The advantage of the partition function theory is in that the generalization of the imputation concept removes the logical difficulty posed earlier, namely that unless the players form a grand coalition, they *cannot* in general expect to get jointly an imputation payoff. Therefore it is natural to put at the basis of the theory the joint payoffs which they *can* get if they are partitioned in a given way. We have seen that this idea underlies also the idea of the bargaining set and of the kernel.

10. N-Person Theory and Two-Person Theory Compared

Usually one speaks of an N-person game only if there are more than two players. It is, of course, also possible to view a Two-person game as a special case of the N-person game (i.e., with $N = 2$). It is instructive to see what happens when we apply the theory so far developed to Two-person games.

The most important distinction in Two-person games is that between constant-sum and non-constant-sum games. This distinction is made also in the N-person game theory. However, in some ways the distinction is sharper in Two-person theory. Let us see why.

First, let us examine the non-cooperative games, i.e., those in which the players may not or cannot (say, for lack of opportunity to communicate) coordinate their strategies. Every Two-person game has equilibria, i.e., pairs of pure or mixed strategies which are characterized by the property that neither player can improve (but may impair) his payoff by departing from an equilibrium-containing strategy while the other keeps to it. In Two-person constant-sum games, all of the equilibrium outcomes will give the same (actual or expected) payoff to each player. Moreover, if each player chooses any of the equilibrium strategies, the outcome will be an equilibrium.

As we have seen in our examples (cf. p. 75), Three-person non-cooperative games also have equilibria, but the payoff vectors may not be the same in all of the equilibrium outcomes. Moreover, it may happen that each player chooses an equilibrium-containing strategy; yet the outcome is not an equilibrium outcome.

It follows that in a Two-person constant-sum game it is always possible to prescribe a choice of strategy: choose a strategy which contains an equilibrium. If both players follow this prescription, an equilibrium outcome will result. In the context of a constant-sum game, such an outcome can be reasonably viewed by rational players as a "solution" of the game.

No such prescription is possible in the N-person game ($N > 2$) if pairs of equilibrium strategies do not result in an equilibrium outcome. The situation in the Two-person non-constant-sum game is similar. Such a game may also contain several equilibria with different payoff vectors, and a choice of an equilibrium-containing strategy by both players may result in an outcome which is not an equilibrium.

Let us now examine the cooperative game. In the case of the N-person game ($N > 2$) we have so far assumed that utilities are transferable and conservative, so that side payments are possible. It therefore makes sense for the players to coordinate their strategies so as to get the largest *joint* outcome. In the context of the Two-person cooperative game, this assumption is usually not made. If it were made, the theory of the Two-person cooperative game would be greatly simplified, as we shall see from the following examples.

Consider the Two-person non-constant-sum game represented by the following matrix.

		Player 2	
		C_1	C_2
Player 1	R_1	3, 4	4, 2
	R_2	1, 3	2, 1

Let us plot the pairs of payoffs as points in a plane with a cordinate system, where the horizontal coordinate represents the payoffs to player 1 and the vertical coordinates the payoffs to player 2. Let us then enclose the points in the smallest convex polygon which includes them inside or on its boundary, as shown in Figure 15.

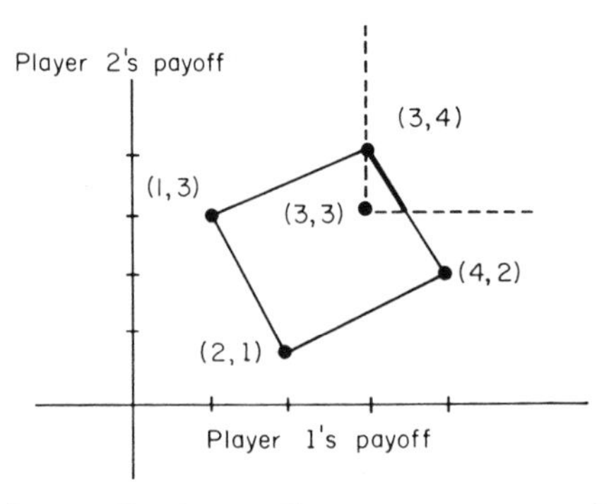

FIG. 15. The payoff polygon. The corners represent the four pure strategy outcomes; the point (3, 3) the security levels of the players; the thick portion of the "northeast" boundary the negotiation set.

The point (3, 3) within this polygon, although not one of the outcomes of the game, has a special significance, because 3 units is what each player can guarantee himself in this game. Player 1 (Row) can be sure of getting at least 3 by choosing R_1, while player 2 (Column) can be sure of getting at least 3 by choosing C_1. Therefore neither player needs to accept less than 3 units and will reject any solution which offers him less than 3.

Now, any point inside or on the boundary of this polygon represents a pair of payoffs which can be realized by appropriate choices of *mixed* strategies. However, if the players cooperate, they need not accept any such pair *inside* the polygon, because any such pair is dominated by at least one pair represented by a point on the line connecting points (3, 4) and (4, 2). The two players, by properly coordinating their (possibly mixed) strategies, can get this pair, which is better for both of them.

It follows that the only pairs of payoffs which the players will accept must lie on the portion of that line intercepted by a horizontal and a vertical line through (3, 3). This line segment is called the *negotiation set* of the game. In fact it constitutes the "basis of negotiations," as diplomats say.

In order to single out a *particular* point on the negotiation set as a solution of the negotiated game, additional assumptions are required; that is to say, additional "criteria of rationality." Here the difference between Two-person theory and N-person theory (as hitherto developed) becomes clear. In the latter, we have assumed that utilities are transferable and conservative and that side payments may be made. As a consequence of this assumption, it is natural to adopt as a criterion of "group rationality" a choice of a point on the negotiation set which gives both (or all) players a maximal *joint* payoff. Such a payoff will also be preferred by both (or all) players, since it will enable them to divide the joint payoff among them so that each of the players can get more regardless of how a smaller joint payoff is divided. Consequently, under this assumption, the outcome of the game ought to be (R_1, C_1). Any other pair of strategies (pure or mixed) yields a joint payoff which is less than the maximal joint payoff.

In Two-person game theory, on the other hand, the assumption of transferable, conservative utility is not usually made. This assumption is avoided because it implies an inter-personal comparison of utilities. Reasons for avoiding such comparisons (as well as for not avoiding them) will be discussed in the last chapter, where we shall examine the philosophical implications of game theory. If utilities of different players cannot be compared with each other, we cannot treat utility as a transferable conservative commodity, hence cannot speak of side payments. For, if one player pays another what is to the giver 1 unit, we do not know how many units the other has received. For the same reason, payoffs cannot be added; and so it makes no sense to say that some particular outcome is associated with a larger joint payoff than another unless both (or all) players get more in one than in another.

We *can* single out outcomes which are not dominated by any other outcomes, in the sense that both (or all) players cannot do individually better in some other outcome. Those non-dominated outcomes which also guarantee to each player at least his security level in fact constitute the negotiation set.

Although, without the possibility of inter-personal comparison of utilities, we cannot narrow down the negotiation set by using the principle of maximizing the joint outcome, there are other ways of narrowing down this set, indeed, to a single point, which then constitutes the solution of the Two-person cooperative game. This, however, can be done in more than one way, and so the solutions singled out by this process are not unique.

Let us examine two methods of arriving at a solution of a Two-person cooperative game (without assuming inter-personal comparisons of utility).[20] One of these we shall call Shapley's procedure, the other the Nash procedure.[21] There is a common idea underlying both methods. First, there is assumed to exist a so-called *status quo* point representing a pair of payoffs which accrues to the players if they fail to agree, or if the "negotiations break down," as the saying goes. For the time being we shall leave the status quo point undetermined. Next, if a solution is found, it is supposed to have the following properties.

1. It must be a point on the negotiation set. Specifically, if it is one of the pairs of payoffs associated with an actual outcome of the game (i.e., one of the points originally plotted in the payoff plane) then it corresponds to a pair of pure strategies. If it is not one of these points, it is on a "northeast" side of the polygon connecting two

such points, and corresponds to choices of mixed strategies. In the latter case, the solution is expressed in terms of expected payoffs.

2. If player 1 assumes the role of player 2, and vice versa the solution must remain the same (except, of course, that the payoffs are interchanged).

3. If the utilities of either or both players independently undergo positive linear transformations (cf. p. 84), the solution must remain the same in the sense of representing the same pair of strategies. (The numerical values of the payoffs will, of course, undergo the same positive linear transformation.)

4. Should the game be enlarged by adding new payoff regions, while the status quo remains the same, then the new solution must either be in the added region or it must remain the same. Mutatis mutandis, if a payoff region not containing the solution is deleted while the status quo point remains the same, then the solution should remain the same.

The first assumption expresses "group rationality" in a context where inter-personal comparison of utilities is not permitted.

The second assumption reflects the irrelevance of the players' personal characteristics to the outcome of the game (a consequence of their "complete rationality").

The third assumption reflects the acceptance of the interval scale as the strongest scale on which utilities can be measured.

The fourth assumption implies independence of the solution from "irrelevant alternatives."

It turns out that if the status quo point is given, then the only point which will satisfy all of these assumptions is uniquely determined. It is, of course, a point on the negotiation set. If the negotiation set is a set of points with coordinates satisfying the equation,

$$y = y(x), \tag{10.1}$$

and if the coordinates of the status quo point are (x_0, y_0), then the only point which satisfies the four assumptions above is the point $[x, y(x)]$ which makes the value of the function

$$f(x) \equiv [x - x_0][y(x) - y_0] \qquad (10.2)$$

a maximum.

That much the two methods—Shapley's procedure and the Nash procedure—have in common. They differ in the choice of the status quo point. In Shapley's procedure, the status quo point is taken to represent the security levels of the players, i.e., what each can be sure to get in the game. Thus, in the game under consideration, the status quo point is $(3, 3)$, because each player can assure himself a payoff of 3. Having fixed the status quo point of our game, we can now "solve" the game by maximizing expression (10.2) with respect to x, after having substituted 3 for x_0, 3 for y_0, and $10 - 2x$ for $y(x)$. The expression achieves a maximum when

$$x = 13/4; y = 7/2. \qquad (10.3)$$

The point $(13/4, 7/2)$ is of course on the negotiation set and is obtained when the two players choose the (coordinated) mixed strategy, in which (R_1, C_1) is chosen with probability 3/4, and (R_1, C_2) with probability 1/4.

The status quo point is determined differently in the Nash procedure. Its determination consists of a choice of so-called "threat strategies." The intersection of these threat strategies is the status quo point. It is assumed that both players commit themselves to the threat strategies in case negotiation fails.

Let us now see how rational players might be expected to choose their threat strategies. There are two considerations. First, should negotiation fail, clearly each player wants the intersection of the threat strategies to be as much in his favor as possible. Hence, player 1 wants x_0 of the status quo point to be as large as possible, and player 2 wants y_0 to be as large as possible. Second, if negotiations lead to an agreement, the status quo point will determine the point on the negotiation set agreed upon. Let us see how.

Differentiating expression (10.2) with respect to x and setting the derivative equal to zero (cf. p. 37), we obtain

$$\frac{d}{dx}(xy - xy_0 - x_0y + x_0y_0) = y + x\frac{dy}{dx} - y_0 - x_0\frac{dy}{dx} = 0 \tag{10.4}$$

$$y = \frac{dy}{dx}(x_0 - x) + y_0. \tag{10.5}$$

Now player 1 (whose point of view we have assumed for the moment) wants y to be as small as possible because, on the *negotiation set,* the interests of the players are opposed: the more one gets in the negotiated solution, the less the other will get. Equation (10.5) shows that on the negotiation set

$$dy/dx < 0; \tag{10.6}$$

that is to say, the slope of the negotiation set line is negative. Consequently player 1, wishing y to be as small as possible, wants x_0 to be as large as possible and y_0 to be as small as possible. Mutatis mutandis, player 2 wants x_0 to be as small as possible and y_0 to be as large as possible. Hence, in the choice of the status quo point, the interests of the two players are diametrically opposed, as if in the choice of that point they were playing a constant-sum game. This constant-sum game is, in fact, easily derived from the original game by (a) "normalizing" the utility scales of the players in such a way that the slope of the portion of the negotiation set which contains the solution equals −1; (b) by constructing a zerosum game in which the payoffs are the differences of the players' payoffs in the original game.

In the game under discussion, normalization leads to the following matrix:

	C_1	C_2
R_1	3, 2	4, 1
R_2	1, $\frac{3}{2}$	2, $\frac{1}{2}$

The difference-of-payoffs game becomes:

	C_1	C_2
R_1	$+1, -1$	$+3, -3$
R_2	$-\frac{1}{2}, +\frac{1}{2}$	$+\frac{3}{2}, -\frac{3}{2}$

We see immediately that both players have a dominating strategy in this game, namely, player 1's is R_1 and player 2's is C_1. That is, in choosing their threat strategies, the players choose the outcome (3, 4) of the original game as the status quo point. Now, substituting 3 for x_0 and 4 for y_0 in expression (10.2), we obtain the expression

$$(x - 3)(6 - 2x). \tag{10.7}$$

The maximum occurs when $x = 3$; consequently $y = 4$. Thus the threat point is also the solution. We see that the Nash procedure arrives at a solution different from that obtained by the Shapley procedure.

Another way of looking at the situation is to apply the Shapley procedure when nothing is known about the game except its characteristic function. Then the problem is how the players are to divide the 7 units which they can jointly get, given that each can get 3 units by himself. In any such division, player 1 will get $3 + a$, and player 2 will get $7 - 3 - a$, where *a* ranges from 0 to 1. But this means that the negotiation set has slope -1 (not -2, which it has in our representation of the game). Applying the Shapley procedure to this scheme, we get the payoff vector (7/2, 7/2), which is the Shapley value solution.

Quite generally, if the security levels of players 1 and 2 are v_1 and v_2 respectively, and they can jointly get c, the assumption of conservative utility states that if one gets $v_1 + a$, the other must get $c - v_1 - a$; so that a ranges from 0 to $c - v_1 - v_2$. The slope of the negotiation set is -1, since $x + y = c$. Applying the Nash bargaining procedure, we maximize

$$(x - v_1)(c - x - v_2) \tag{10.8}$$

and obtain

$$x = 1/2(c + v_1 - v_2) \tag{10.9}$$

$$y = 1/2(c - v_1 + v_2). \tag{10.10}$$

But this is also the Shapley value solution, as the reader can verify.

We see, then, that three different ways of viewing the cooperative game leads to three different solutions. It is instructive to see why this is so.

First of all, it is important to realize that the "discrepancy" between the Shapley value solution (7/2, 7/2) and the Shapley procedure solution (3.25, 3.50) is not really a meaningful discrepancy if the payoffs are given on an interval scale. It would be wrong, for example, to conclude that in the Shapley procedure the players do not get jointly as much as they get in the Shapley value solution (namely 6.75 in the latter compared with 7 in the former). One must keep in mind that the "sum of the payoffs" has no meaning whatsoever in the context of the Shapley procedure as we have described it. This method does not answer the question "How much will the players get jointly and how will they divide the joint payoff among them?" It asks and answers the question "How should the players coordinate their strategies if the solution is to satisfy certain criteria of rationality (or equity)?" When the proper coordinated strategy (pure or mixed) is obtained, each player gets a certain (actual or expected) payoff. But the two payoffs of the players cannot be compared with each other. The payoff received by each player can only be compared with other payoffs which he might have received.

The Shapley value approach assumes more about the payoffs than the Shapley procedure (namely that they can be added, etc.). At the same time both the Shapley value solution and the Shapley procedure use less information than is available about the game itself. They make no reference to the strategies, only to the security levels of the individual players and what they can get on the negotiation set. The Nash procedure uses more information about the game, because in the process of singling out the "best" strategies available to the players it examines not only the strategies associated with the negotiation set, but all of them. Thus the bargaining leverage which the players derive *from what can actually happen* in the game is reflected in the Nash procedure but not in the Shapley procedure or in the Shapley value. For this reason some authors (e.g., Luce and Raiffa) call procedures such as Shapley's (there are other such procedures) *arbitration*, as if the outcome of the game were determined by a third party on the basis of some equity principle, and procedures like Nash's *bargaining*. It is assumed that in the process of bargaining, the players can indicate to each other what they are in a position to do to each other if agreement is not reached. An arbitrator interested in satisfying some equity principle is likely to give less weight to such arguments.

The characteristic function approach to N-person games posits transferable, conservative (hence additive) utilities. If this approach is projected to the Two-person game, an extremely simple theory results, because so much information is thrown away about the actual structure of the games. All constant-sum Two-person games have the characteristic function

$$v(\bar{1}) = v_1;\ v(\bar{2}) = c - v_1;\ v(\overline{12}) = c. \qquad (10.11)$$

Since $v(\bar{1}) + v(\bar{2}) = v(\overline{12})$, all such games are inessential and present no interest in their characteristic function form.

As for the non-constant-sum games, we have seen that their characteristic functions can be normalized so as to have the form

$$v(\bar{1}) = 0;\ v(\bar{2}) = 0;\ v(\overline{12}) = 1. \qquad (10.12)$$

Therefore all such games, viewed from the point of view of the characteristic function, reflect a situation in which two players have some unit of "good" to divide among them, it being understood that if they fail to agree on the division, neither gets anything. From (10.12) we can only infer that the Shapley value of *every* normalized non-constant-sum Two-person game is (1/2, 1/2).

There is, of course, much more to be said about Two-person games. But the characteristic function representation is not able to bring it out. For this reason it is not interesting to treat Two-person games in the context of characteristic functions. In the context of the N-person game ($N > 2$), all of the information pertaining to the actual strategic structure of the game has likewise been thrown away. However, enough complexity *remains* when the games are presented in characteristic function form to provide material for a complex theory. It is well to keep in mind, however, that a tremendous amount is left unexamined and unexplored. In the next chapter, we shall examine an extension of the Nash procedure to the general N-person game.

11. *Harsanyi's Bargaining Model*

In the preceding chapter we compared two procedures for arriving at a solution of a cooperative Two-person non-constant-sum game, namely Shapley's and Nash's. We have seen that if Shapley's procedure is applied under the assumption of transferable utility, it reduces to the Shapley value solution of the Two-person game. If transferable utility is not assumed, then Shapley's procedure amounts to choosing the security levels of the two players as the status quo point and applying the so-called Nash-Zeuthen principle[22] to obtain the solution point on the negotiation set. This procedure can be defended on the grounds that the status quo point of a bargaining situation can be reasonably taken as the pair of payoffs which the players can be respectively sure of obtaining. However, it ignores the strategies available to the players in the actual game. These strategies are taken into account by the Nash procedure. For this reason the Nash procedure can be considered more relevant to game theory in that context.

Harsanyi's bargaining model is essentially an extension of the Nash procedure to the N-person game ($N \geqslant 3$). According to Harsanyi, this model, too, arrives at a unique payoff disbursement, which can be recommended to the (rational) players of an N-person game, in the sense that it takes into account the bargaining positions of the players derived from the strategies available to them. The added feature is that not only the bargaining positions of the *individual* players are taken into account (as in Nash's procedure applied to the Two-person game) but also the bargaining positions of all possible *subsets* of players. These subsets are called *syndicates.*

They are distinguished from coalitions in that each player is considered as a member of every syndicate viewed as a set of players which includes him. Actually, the syndicates can be viewed as the *potential* coalitions. They enter N-person game theory in other contexts also, for example, in the definition of the core (cf. Chapter 4), of ψ-stability (cf. Chapter 8), of the bargaining set (Chapter 6) and of the kernel (Chapter 7).

Consider a particular pair of players, i and j. Ignore for the moment what all the other players can do; that is, imagine that the payoffs of all the other players have been fixed, while we examine the relative positions of i and j. As in the Nash procedure, assume that there is a *disagreement payoff vector*

$$\vec{d} \equiv (d_1, d_2, \ldots, d_n), \tag{11.1}$$

a point in the payoff space P. This payoff vector will obtain if even one of the n players fails to cooperate in achieving a Pareto-optimal payoff.

Ignoring the other players, and applying the Nash procedure to the two players, i and j, we obtain *their* payoffs in the solution by maximizing the product of their gains with respect to the corresponding components of $\vec{d}$ (cf. Chapter 10), namely the product

$$\pi_{ij} = (x_i - d_i)(x_j - d_j). \tag{11.2}$$

The maximization is subject to the following conditions:

(i) The resulting payoff vector $\vec{x} \equiv (x_1, x_2, \ldots, x_n)$ shall be in the payoff space P.

(ii) $x_i \geqslant d_i$; $x_j \geqslant d_j$.

(iii) $x_k =$ constant for $k \neq i, j$.

If now we impose these conditions on all the pairs (i,j), this amounts to maximizing the product

$$\prod_{i=1}^{n} (x_i - d_i), \tag{11.3}$$

subject to the conditions

$$\vec{x} \in P; x_i \geqslant d_i. \quad (11.4)$$

The mathematical properties of the payoff space P (compactness, convexity) are such that a unique vector $\vec{x}^*$, satisfying the above conditions, can always be found except in some uninteresting degenerate cases.

One could also arrive at this result by imposing the axioms of the Nash procedure to the situation with N players, namely:

1. *Collective rationality.* If $\vec{u} \equiv (u_1, u_2, \ldots, u_n)$ is a solution, there is no point in the payoff space which accords to *every* player a larger payoff than in $\vec{u}$.

2. *Symmetry.* If the game is symmetric, every player ought to get the same payoff (since the players are indistinguishable).

3. *Linear invariance* (cf. p. 84).

4. *Independence from irrelevant alternatives* (cf. p. 163).

So far, the proposed solution of the N-person game is a straightforward generalization of the solution arrived at by the Nash-Zeuthen principle. That is, we have assumed that the status quo vector is *given.* Furthermore, we have assumed that only a pair of players at a time participate in the bargaining, while all the others stay put, as it were. To effect a full generalization, we must drop both assumptions. The players do not know a priori what the disagreement payoff vector will be. Nor do the remaining players remain silent on the side lines, as two of them are bargaining. On the contrary, they may well interpose remarks and make offers which are quite relevant to what the two are bargaining about. In fact, we must suppose that every subset of players, at least temporarily, acts like a coalition in the bargaining process. This is the meaning of the syndicate.

The outcome of the game will be determined, to quote

Harsanyi, "by a whole network of $(2^n - 1)$ agreements concluded by the members of different syndicates."

Let us now suppose that every syndicate guarantees to its members certain payoffs, which shall be called *dividends.* The members of the syndicate agree (unanimously) to pool their efforts to secure the dividends for its members. A player is a member of several such syndicates. He expects dividends from each of them and expects these dividends to be additive.[23] It should be noted that some such dividends may be negative. For example, the members of a particular syndicate may know that, as a result of the outcome of the game, they will get negative payoffs if they act like a coalition. Consequently, if they want their proposed dividends to be *guaranteed,* they can declare only negative dividends. The total dividends accruing to a player will constitute his final payoff. For completeness, we shall assume that dividends accrue to a player even from syndicates to which he does not belong. These will, of course, be zero. This assumption will serve to simplify our notation, since by indicating sums of dividends we can sum over all the subsets of a certain class instead of specifying each time the subsets to which the player in question belongs.

Now, in the process of bargaining each syndicate S has at its disposal a *threat strategy* $\theta(S)$. The syndicate announces (publicly) that they will resort to this strategy if no agreement can be reached between the members of S and the members of $-S$, as to the payoffs which the members of S will receive. If the members of both the syndicates S and those of the complementary syndicate $-S$ were to resort to their respective threat strategies $\theta(S)$ and $\theta(-S)$, an outcome of the game would result, and consequently a payoff to each player, namely the vector $c(S)$, whose components are either $c_i(S)$, if player i belongs to S, or $c_j(-S)$ if player j belongs to $-S$.

Now if the players are rational, all these publicly stated commitments must be consistent with each other.

In particular, if the dividends are *guaranteed,* the total dividends promised to a player by *subsets* of a given syndicate to which he belongs cannot exceed the payoff which accrues to him should his syndicate have to resort to its threat strategy (S). Hence if R is any subset of S, we must write

$$\sum_{R \subseteq S} w_i(R) \leqslant c_i(S), \qquad (11.5)$$

where $w_i(R)$ is the dividend promised to i by the syndicate R. Note that it does not matter whether i belongs to any particular R or not, since the dividends from the subsets to which he does not belong enter the sum as zeros.

The syndicate composed of all the n players we shall call the *grand syndicate,* and all the smaller ones *sectional syndicates.* When all the sectional syndicates have announced the guaranteed dividends and the threat strategies, the grand syndicate may also agree to distribute its dividends, $w_i(N)$, among its members. Clearly, the members of the grand syndicate need not announce any threat strategy, since there is no complementary syndicate to "threaten." If the final payoff to each player is to equal the cumulated dividends from all the syndicates, we must have

$$u_i = \sum_{R \subseteq N} w_i(R), \ \vec{u} \in P. \qquad (11.6)$$

Now the fact that some syndicate may agree to pay negative dividends to certain of its members makes it possible to forgo the restriction, which is always in force with regard to the actual payoff of a game, namely that the payoff vectors must be in the payoff space. We may in fact have dividends assigned in such a way that the vector sums $\sum_{R \subset N} w(R)$ are vectors outside the payoff space P (i.e., may not be realizable); and also we may have

$$\sum_{R \subset S} w_i(R) > c_i(S), \quad (11.7)$$

which is to say that the dividends accruing to a member of S through the dividends declared by the *proper* subsets of S may exceed the payoffs which accrue to him if the syndicate S must face a showdown with –S. To make these results consistent, we must make proper adjustments in the dividends declared by N or by S (the supersets of R). This will actually imply that such dividends will be negative at least for some of the members. In such cases, the dividends $w_i(R)$ will be called *conditional* dividends. That is, they are conditional on the supposition that the members of the larger syndicates (those that include the syndicate R) will agree among themselves on negative dividends with regard to at least some of the members.

To see the significance of this, imagine two players i and j as one-person syndicates. Each has a threat strategy, which, if resorted to (against the coalition of all the others), will bring $c_i(I)$ and $c_j(J)$ respectively. Each player can guarantee himself a dividend larger than $c_i(I)$ or $c_j(J)$ *on condition* that the two of them agree to take a loss if they have to join in a coalition against all others and face a showdown.

Harsanyi envisages the bargaining process in three stages. First, each of the syndicates (except the grand syndicate) announces its threat strategy. These determine the conflict payoffs to every sectional syndicate. Next the sectional syndicates announce their dividends consistent with (11.5). Finally the grand syndicate announces its dividends in accordance with (11.6).

We must also require the following rule governing the cases where some sets of players are unable to agree on either the threat strategy they will use or on the dividend vector to be apportioned among them. If they cannot agree, then the dividend declared by the syndicate S to each member of S shall be zero, and moreover the dividends from all the *supersets* of S shall be zero to the members of S. The idea is that, if these players are unable to agree among themselves, they are ipso facto prevented from coordinating their strategies in any larger coalition if the game is to be played. Hence they make all the coalitions which include them inoperative.

Harsanyi shows how this procedure leads to a unique outcome of the game, in the sense of a disbursement of payoffs. The reasoning is quite involved. Rather than pursue it in all the ramifications, we shall demonstrate the process by an elementary example. The game shall be the Three-person Prisoner's Dilemma game discussed in Chapter 2, under the condition that the moves of the players are concealed. That is, the game is now *not* a game of perfect information. Under this condition, each of the players has just two strategies, C and D. The eight possible outcomes of the game are as follows:

Strategy triple			Payoff vector		
1	2	3	1	2	3
C	C	C	1	1	1
C	C	D	0	0	3
C	D	C	0	3	0
D	C	C	3	0	0
D	D	C	2	2	−2
D	C	D	2	−2	2
C	D	D	−2	2	2
D	D	D	−1	−1	−1

Assume first that no *conditional* dividends are announced. Thus, every dividend declaration must be backed by what the syndicate will get in case of a showdown. Consider first the two-player syndicates. They view the process of choosing a threat strategy against the third player as a Two-person non-cooperative game, shown in the following matrix, in which the entries are the payoff vectors.

		Third Player's Strategies C	Third Player's Strategies D
Syndicate's Strategies	CC	1, 1, 1	0, 0, 3
	CD	0, 3, 0	−2, 2, 2
	DC	3, 0, 0	2, −2, 2
	DD	2, 2, −2	−1, −1, −1

In this game the third player's strategy D dominates C. Being rational, he is expected to choose it. Now the syndicate can eliminate its strategy (DD), since it is dominated by strategy (CC). The positions of players 1 and 2 are symmetric. Consequently they can agree either on (CC) or on a 50–50 *mixture* between CD and DC. In either case, the expected payoff to each player of the pair syndicate is zero. Consequently the syndicate can guarantee only dividend 0 to each of its members.

Let us now look at the situation of the third player. Whatever the syndicate does (rationally), he can expect at least 2. But can he declare a dividend of 2 for himself? Not if the dividends are to be *unconditionally* guaranteed. For he must also consider the situation when *he* is in a two-player syndicate, which can only guarantee 0 to its members. Consequently, player 3 cannot declare an (unconditional) dividend of 2 for himself. The most he can guarantee himself is 0.

All the single players and pairs having announced their dividends, it remains for the grand syndicate to announce its. Clearly, this syndicate can get 1 for each of its members; and this is what each will get, namely the payoff of the cooperative outcome (CCC).

Now let us admit the possibility of conditional dividends. It makes sense now for each single player to "claim" 2. To make this claim consistent with what a two-person syndicate can announce, the player must agree to a dividend of −2 from the pair syndicate. He can refuse to agree to this, which cancels the dividends of the pair syndicates *and* of the grand syndicate, since the latter includes the pair syndicate. The single player does not care, for he is still ahead with his claim of 2. Now, it turns out that if each of the players makes this claim, then several points of the payoff appear as possible solutions of this game. We seek, however, a unique solution. If a situation of this sort occurs, Harsanyi proposes *further* bargaining among the players, whereby the disagreement vector is now given as the vector that awards to each player the minimal payoff of all the payoffs in the set which constitutes the solution. Because of the symmetry of this game, such a situation leads once more to the final outcome, with payoff vector (1, 1, 1). If the game were not symmetric, the differences in the bargaining positions of the players would be reflected in the final outcome.

It appears, then, that Harsanyi's bargaining model is an adaptation of Nash's procedure to the N-person game. But while it is based on the same principles, its "logic" is much more complex, because of the intricate ramifications that result from considering simultaneously (in counterpoint, as it were) the bargaining positions of all the interlocking syndicates.

The theory is considerably simplified if it is developed in the context of transferable utilities.

If the utilities are transferable and additive, we can speak of a joint payoff which accrues to every coalition S in any given outcome. Now when a syndicate S announces a threat strategy (and the counter-syndicate announces its counter-strategy), the syndicates are determining a possible outcome (in case the threat strategies should be realized) which will award the joint payoff $\alpha[\theta(S), \theta(-S)]$ to S and $\alpha[\theta(-S), \theta(S)]$ to $-S$. The choice of threat strategy can now be more flexible, since the possibility of reapportioning the payoffs gives the syndicate a wider latency in coming to an agreement. Threat strategies, which might have been vetoed by members who stand to lose much in a showdown, may now be acceptable if the other members of the syndicate promise to compensate the losers.

As in the case of non-transferable utilities, the purpose of these threat strategies is to establish a showdown outcome, to be used as a basis for arbitrating the final outcome. Each syndicate tries to establish this showdown outcome so as to maximize its own final return (which depends on the showdown outcome). This maximization is accomplished by making the difference between the showdown payoffs of the two syndicates as large as possible in favor of one's own. But this difference, if viewed alternatively from the standpoint of a syndicate and the counter-syndicate, differs only in sign. Therefore, as the two syndicates parry for a choice of threat strategies, they are essentially playing a Two-person constant-sum game (quite regardless of whether or not the original game was constant-sum). Now, the solution of a constant-sum Two-person game is known. It is obtained when each side finds its (pure or mixed) maximin strategy, which turns out to be also its minimax strategy, the maximization being of one's own payoff under the restriction that the other player attempts to minimize it. In symbols, we express the showdown payoffs to every syndicate S by the following function:

$$v'(S) = \underset{\theta(S)}{\text{Max}}\ \underset{\theta(-S)}{\text{Min}}\ \alpha[\theta(S), \theta(-S)]. \qquad (11.8)$$

The function $v'(S)$ can be shown to be a suitable characteristic function.

Imagine now that the players are playing a cooperative N-person game G′ with characteristic function $v'(S)$ derived in the manner described from the original characteristic function $v(S)$ of the original game G. The Shapley value solution of G′ constitutes Harsanyi's solution of G. This solution differs from the Shapley value solution of the original game in that, by transforming the characteristic function, it takes into account not only the characteristic function of the game but also the strategies available to the players and to all the subsets of players in the original game.

One might ask in this connection why the earlier models of the equilibrium properties of the N-person game (except the Shapley value) failed to come up with unique payoff vectors as solutions. I believe that this is because Harsanyi's conception of "rationality" is stronger than in the original formulation of game theory. Harsanyi's players, in pursuing their bargaining strategies, see further than, say, the Von Neumann-Morgenstern players. A Harsanyi player imagines himself not simply as a member of a particular coalition which has formed, or is about to form, but rather as a member of every possible coalition which can form, and he does so simultaneously. Thus, if it appears to a player that by joining with someone he can get more than he does in a grand coalition, he must consider also what would happen if he had to join with those other players whom he plans to "double cross." Recall that this is also the procedure in arriving at the Shapley value of the game, where each player calculates his "prospects" in every possible coalition which can form, including the order in which the players join it. Similar calculations underlie the bargaining strategies of Harsanyi's players, plus the opportunities made apparent by the strategic structure of the original game, not utilized in Shapley's model.

The Harsanyi solution represents, perhaps, the farthest advance into a theory of bargaining that can be made by purely game-theoretic methods. That is to say, the process of arriving at the solution utilizes all the information about an N-person game that can possibly be utilized, and arrives at a unique result of each game.[24] Any further development of the theory will have to "breach the psychological barrier," as it were; i.e., take into account special psychological (and other non-strategic) aspects of conflicts of interest.[25]

PART II. *Applications*

Introduction to Part II

In Part II, we shall be concerned with "applications" of the essential ideas of N-person game theory discussed in Part I. The quotation marks are meant to forestall misunderstandings. When we speak of "applied science," we often refer to techniques derived from the understanding of general principles and directed toward controlling some part of our environment. This conception hardly fits the "applications" of game theory to be presently discussed. To begin with, the gap between this theory and life situations is far greater than that between the abstractions of natural science and the events from which they are derived. Even if this gap could be narrowed, it is still doubtful whether the "applications" of game theory could become direct analogues of the applications of natural science. Clearly if game theory has any relevance to matters outside of pure mathematics, it is to matters of concern to the social scientist. In the social sciences, the objects of study (men and institutions) are also subjects, that is, choosers of alternatives. Whereas nature may remain passive while we study and manipulate it (even this is true only within limits), men do not. It is therefore unrealistic to envisage a future "applied social science" as an essential replication of applied natural science, unless by "social control" one means self-control, where subject and object are identical.

A prerequisite of self-control is understanding. This brings us to the protracted controversy concerning the objectives of social science: whether "understanding" (*das Verstehen*) of social events can be an objective per se, as is sometimes asserted by those social scientists who reject the strict epistemological criteria of the positivists and the circumscribed utilitarian criteria of the pragmatists.

Avoiding this controversy for the present with regard to the general objectives of science, I should like to reiterate my conviction that the understanding of the logical structure of strategic conflicts is indeed the prime and, at least at present, the only achievable objective of game theory. However, "understanding" in this context is not the intuitive understanding sought after by the social scientist of the old (pre-positivist) school, nor the understanding of the positivist (rigidly linked to the ability to predict and to control). It is rather the understanding of the mathematician. Based on most rigorous analysis, it is *impersonal* (hence has a partial claim to scientific validity); but it is also independent of the ability to predict or to control (unlike the understanding imparted by sciences with empirical content). The conclusion of a mathematical theorem predicts nothing except that any competent mathematician must come to the same conclusion if he starts from the same hypotheses.

The mathematician attains "understanding" by gaining an insight into the interdependence of logical relations. Similarly, a game theoretician attains an "understanding" of the strategic components of conflict situations by gaining an insight into the often extremely intricate interrelations of strategic considerations. This insight does not reveal techniques for "controlling" conflicts (let alone means of "winning" them); but gaining it is an important step forward—toward what we do not know, because we do not know what new goals may be revealed by the increased understanding. Nevertheless it is possible to defend the view that *any* gain of understanding of matters that have some theoretical bearing on important aspects of human relations is a proper human goal.

Accordingly, the term "applications" will be justified by translating the purely mathematical concepts into highly simplified and idealized, but nonetheless imaginable, social situations, such as "a small market" (two buyers and a seller), "a large market" (of many traders), "a legislature" (a voting body defined exclusively by its decision rules), "a committee" (similarly defined), "a small group competing individually and in coalitions for advantages strictly in accordance with prescribed rules" (a gaming experiment), etc. The corresponding situation will be analyzed in terms of the concepts described in Part I, and purely logical conclusions will be drawn from the analysis. These conclusions, it is hoped, will give the reader a gratifying feeling of seeing the depicted situations in a new light. In the instances involving controlled experiments (in some cases also field observations), comparisons will be made between the conclusions drawn and the observed results. As usual, agreement between theoretically derived conclusions and results will be offered as a partial corroboration of the relevant theoretical models. Discrepancies will be viewed as occasions for raising further probing questions concerning matters not taken into account in the models.

All this is part of the standard operating procedure in "hard" experimental sciences. However, to view game theory as no more than a theoretical framework of an experimental science would be to miss its more important significance as a vehicle of a very special understanding it imparts, quite aside from any connections we may be able to establish between its essential ideas and human behavior.

12. *A Small Market*

A market consisting of one seller and two buyers can be represented by a Three-person game in characteristic function form. Suppose the seller (player 1) has a house for sale, his "rock bottom price" being $20,000. If the other two players are to be bona fide buyers, each ought to be willing to pay at least $20,000. Suppose one (player 2) is willing to pay up to $25,000, while the other (player 3) is willing to pay up to $27,000. Can we say anything about what will transpire? If the two buyers lay their cards on the table, as it were (that is, if they declare their maximum offers respectively), then clearly the house will go to the higher bidder for his top price. On the other hand, if only the seller declares the minimum sale price, players 1 and 2 may bid against each other until the bid exceeds $25,000, at which time the house will go to player 3 for a price just in excess of that amount. Or one of the players may acquire the house for the minimum price and compensate the other for letting him do so.

In practice, this seldom happens. All three keep their minima or maxima to themselves, the seller in the hope of getting as much as possible, the buyers in the hope of getting the house as cheaply as possible.

Let us see what we can conclude if we represent the situation formally as a Three-person game.

Passing to the general case, assume that some indivisible object is worth a units to player 1, b units to player 2, and c units to player 3. (Without loss of generality, we can assume $a \leqslant b \leqslant c$.) Player 1, being the seller, has the object in his possession. Since he is not forced to give up the object, nor pay anything to either other player, he is assured of a payoff of a units. On the other hand, he is not sure of getting more since neither player is forced to buy. Therefore a is his "security level" in this game or, in terms of the characteristic function of the game, $v(\overline{1}) = a$. As for players 2 and 3, neither of them can be sure of getting the object nor are they forced to pay anything. Their security levels are therefore both 0, so that $v(\overline{2}) = 0$; $v(\overline{3}) = 0$.[26]

Let us now see what the possible coalitions can assure themselves. If players 1 and 2 join in a coalition, they have the object in their possession, hence can assure themselves b units, which is what the object is worth to player 2. Similarly, if players 1 and 3 join in a coalition, they can assure themselves c units. Players 2 and 3 can assure themselves nothing even in a coalition (since player 1 need not sell). In summary, $v(\overline{12}) = b$; $v(\overline{13}) = c$; $v(\overline{23}) = 0$. All of them together have the object, which is worth c to player 3; hence $v(\overline{123}) = c$.

An imputation in this game is any payoff vector (x_1, x_2, x_3) in which

$$x_1 \geqslant a,\ x_2 \geqslant 0,\ x_3 \geqslant 0,\ x_1 + x_2 + x_3 = c. \quad (12.1)$$

Let us see what the outcome will be if it is one of the outcomes in the core of the game. Recall that these imputations are such that no two players can get more jointly (than they get in the imputation) if they should form a coalition against the third. This condition is expressed by the following inequalities:

$$x_1 + x_2 \geqslant b \quad (12.2)$$

$$x_1 + x_3 \geqslant c \quad (12.3)$$

$$x_2 + x_3 \geqslant 0. \quad (12.4)$$

Solving the system of inequalities (12.2)–(12.4) together with (12.1) we obtain the imputations of the core, namely

$$b \leqslant x_1 < c;\ x_2 = 0;\ x_3 = c - x_1, \quad (12.5)$$

which is to say that the object will go to player 3 for some amount between b and c. This would be the case, for example, if player 2 bid and player 1 rejected this bid, having found out that player 3 will give more, but not how much more.

While this set of outcomes seems in accord with what one might expect on common sense grounds, it fails to reflect the possibility of the two buyers forming a coalition for the express purpose of forcing player 1 to sell the object below b. This is clearly possible, since the seller will accept as little as $a \leqslant b$. To bring this about, the buyers must agree between themselves to keep the bids down. This is not reflected in the core, but the Von Neumann-Morgenstern solution takes this possibility into account.

Consider the set of imputations consisting of all the imputations in the core given by (12.5) plus all the imputations of the form

$$J = \{x, 1/2(c - x), 1/2(c - x)\}, \text{ where } a < x < b. \quad (12.6)$$

This set forms a solution. Clearly no imputation in the core dominates any other imputation in the core nor is dominated by any other imputation (by definition of the core).

Consider now any imputation not in the set defined by (12.6), say

$$\vec{x} \equiv [x_1, t(c - x_1), (1 - t)(c - x_1)], \quad (12.7)$$

where $0 \leqslant t \leqslant 1$; $t \neq 1/2$. We must show that one of the imputations belonging to the set J dominates $\vec{x}$. Suppose first that $x_1 < b$. Then we can choose $x > x_1$, say $x_1 + \varepsilon < b$, where ε is an arbitrarily small positive number. This singles out an imputation

$$x' \equiv [x_1 + \varepsilon, 1/2(c - x_1 - \varepsilon), 1/2(c - x_1 - \varepsilon)], \quad (12.8)$$

from the set of J. Suppose $t < 1/2$ (otherwise, take $1 - t$). Then we can choose an ε so small that

$$1/2(c - x_1 - \varepsilon) > t(c - x_1), \qquad (12.9)$$

which will be the case if $\varepsilon < (1 - 2t)(c - x_1)$. Then $\vec{x}'$ dominates $\vec{x}$ via $(\overline{12})$.

If on the other hand $b \leqslant x_1 < c$, we choose an imputation from the core, with $x = x_1 + \varepsilon$, say

$$\vec{x} \equiv (x_1 + \varepsilon, 0, c - x_1 - \varepsilon). \qquad (12.10)$$

If $t > 0$, we can again choose an ε so small that

$$c - x_1 - \varepsilon > (1 - t)(c - x_1). \qquad (12.11)$$

Now $\vec{x}'$ dominates $\vec{x}$ via $(\overline{13})$. Note that if $t = 0$, $b \leqslant x_1 < c$, then $\vec{x}$ is in the core, hence in the solution.

This solution, therefore, comprises all the outcomes where either (1) the object goes to player 3 for some price x, between *b* and *c* (in the case of an imputation in the core); or (2) the object is sold to player 3 for some price x between *a* and *b*, whereupon player 3 pays off to player 2 one half of what he has saved, i.e., $1/2(c - x)$.

The transaction between the two buyers represents their apportionment of the savings they were able to effect jointly (buying the house for less than *c*) by forming a coalition, i.e., not bidding against each other.

The solution seems to be a "sensible" one. Were it the only solution to this game, we would say that game theory provides a corroboration of what is likely to be observed. Unfortunately, the game has other "solutions"; i.e., outcomes which satisfy the requirement of the Von Neumann-Morgenstern solution.

Consider any two monotonically decreasing functions of x: $f(x)$ and $g(x)$, such that $f(x) \geqslant 0$, $g(x) \geqslant 0$, and $x + f(x) + g(x) = c$. Examples: $f(x) = c_1 - x/2$; $g(x) = c_2 - x/2$, where $c_1 + c_2 = c$ and $f(x)$ is defined for $x \leqslant \text{Min}\,[2c_1, 2c_2]$.

Then the core plus all the triples $(x, f(x), g(x))$,

where $a \leqslant x \leqslant b$ also form a solution.[27] Note that if *any* such triple is in a solution, this says no more than that the object will go for some price between *a* and *b*, and the two buyers will make *some* adjustment so as to apportion the amount saved. If we add the imputations of the core, we admit the possibility that the object will go (to player 3) for some price between *b* and *c*, player 2 being left out of the deal. In other words, every conceivable outcome which does not violate common sense is included in some solution, and game theory seems to say no more than some such common sense-dictated outcome can be expected to occur. Is this so?

It can be argued that the Von Neumann-Morgenstern theory says a little more. In the light of this theory, all the common sense outcomes can be partitioned into sets, each set constituting a solution. Thus a particular common sense outcome, although it is in some solution, is not in all the solutions. This raises the question of whether in some particular situation we may not observe outcomes all belonging, say, to a single solution. Or, to put it in another way, whether we cannot associate with each solution some "standard of behavior" which, being beyond the scope of game theory to establish, must be assumed as given if a particular solution is to be singled out.

This we can easily do. We shall first observe that the core is a subset of all the solutions of this game, and expresses the following principle: if the seller knows who is the low bidder and what his top price is, he negotiates only with the higher bidder (by hoisting up his "bottom price" to what he can always get from the low bidder).

The distinguishing feature of any solution is contained in the functions of $f(x)$ and $g(x)$. The specific form of these functions tells us about the "standard" governing the division of the savings which the two players are able to effect if they make a deal not to reveal their "top offers."

Of course we can deduce such standards only if we are able to observe several transactions consistent with some fixed functions f(x) and g(x). If so, the Von Neumann-Morgenstern solution provides us with some theoretical leverage to explain the behavior of parties in a bargaining situation.

Let us now examine the Shapley value solution of the same game.

For the general Three-person game, the Shapley value solution is

$$\phi_1 = \tfrac{1}{3}[v(\overline{123}) - v(\overline{23})] + \tfrac{1}{6}[v(\overline{12}) - v(\overline{2})] + \tfrac{1}{6}[v(\overline{13}) - v(\overline{3})] + \tfrac{1}{3}v(\overline{1}) \quad (12.12)$$

$$\phi_2 = \tfrac{1}{3}[v(\overline{123}) - v(\overline{13})] + \tfrac{1}{6}[v(\overline{12}) - v(\overline{1})] + \tfrac{1}{6}[v(\overline{23}) - v(\overline{3})] + \tfrac{1}{3}v(\overline{2}) \quad (12.13)$$

$$\phi_3 = \tfrac{1}{3}[v(\overline{123}) - v(\overline{12})] + \tfrac{1}{6}[v(\overline{13}) - v(\overline{1})] + \tfrac{1}{6}[v(\overline{23}) - v(\overline{2})] + \tfrac{1}{3}v(\overline{3}). \quad (12.14)$$

Substituting the values given by the characteristic function, we have

$$\phi_1 = \tfrac{1}{3}(c - 0) + \tfrac{1}{6}(b - 0) + \tfrac{1}{6}(c - 0) + \tfrac{1}{3}(a) = \tfrac{1}{6}(2a + b + 3c) \quad (12.15)$$

$$\phi_2 = \tfrac{1}{3}(c - c) + \tfrac{1}{6}(b - a) + \tfrac{1}{6}(0 - 0) + \tfrac{1}{3}(0) = \tfrac{1}{6}(b - a) \quad (12.16)$$

$$\phi_3 = \tfrac{1}{3}(c - b) + \tfrac{1}{6}(c - a) + \tfrac{1}{6}(0 - 0) + \tfrac{1}{3}(0) = \frac{3c - 2b - a}{6}. \quad (12.17)$$

In the context of our example (sale of house), this means that the value of the game to the seller is $24,-333.33, to the low bidder $833.33, and to the high bidder $1,833.33. To realize this outcome, the house should be sold to the high bidder for $24,333.33, who will then "compensate" the low bidder (out of his effected "saving" of $2,666.67) in the amount of $833.33, i.e., one-third of the saving.

Comparing ϕ_1, ϕ_2, and ϕ_3 in the general case, it is interesting to observe that the seller gains most and the low bidder least in all cases, regardless of the values of *a*, *b*, and *c*, except that $a \leqslant b \leqslant c$ (which we assume is always true in a market of this sort). Moreover, the amount gained by the low bidder depends only on the difference between his top offer and the seller's bottom price, not on the high bidder's top offer. This last result is not a priori obvious. It is in some way a consequence of the special assumptions which lead to the Shapley value solution.

Examining the game from the point of view of ψ-stability, we must specify some restriction on the sort of changes which a given coalition structure can undergo. However, in the case of a Three-person game, the imposition of such restrictions tends to trivialize the problem. Suppose, for example, the coalition structure under consideration is $\{(\bar{1}), (\bar{2}), (\bar{3})\}$; i.e., each player is alone. To prohibit some coalition, say $(\overline{23})$, reduces the bargaining to a contest between players 2 and 3 for the cooperation of 1. Nothing is left but for the seller to choose the more profitable alliance. A similar situation obtains with regard to other coalition structures. Recall that the case of *no* restrictions on coalition formation is also subsumed under the ψ-stability theory. Let us therefore assume that there are no such restrictions and examine the corresponding stable pairs $\{\vec{x}, \mathfrak{J}\}$, where $\vec{x}$ is an imputation and $\mathfrak{J}$ is a coalition structure.

We find that if $\mathfrak{J} = \{(1), (2), (3)\}$, then the only imputations which form stable pairs with this coalition structure are those where

$$b \leqslant x_1 < c;\ x_2 = 0,\ x_3 = c - x_1. \qquad (12.18)$$

These are the imputations of the core, except for the single imputation (c, 0, 0). This one is excluded by the requirement of the ψ-stability theory that, to participate in a coalition with another, a player must have some positive inducement to do so. Similarly we find that if $\mathfrak{T} = \{(\overline{13}), 2\}$, i.e., if the seller is in coalition with the higher bidder, then the imputations which form stable pairs with it are also essentially those of the core.

As for the other two possible coalition structures, namely $\{(\overline{12}), (\overline{3})\}$ and $\{(\overline{1}), (\overline{23})\}$, no imputations form stable pairs with either.

It turns out, then, that the only stable imputations are those of the core, and the only stable coalitions are those where each player is alone, and where the seller is in a coalition with the higher bidder.

Thus the ψ-stable pairs exclude many imputations from the class which are among the "stable" outcomes (if the imputations in the solution are thought to have this property), something which the Von Neumann-Morgenstern solution does not do. It is not hard to see the reason for this exclusion, namely the inherent instability of the coalition between the two buyers, each of whom is susceptible to tempting offers from the seller, who can play one against the other. If the two buyers make a binding agreement with each other not to accept overtures from the seller, another situation ensues. For this is equivalent to prohibiting any changes in the coalition structure if the coalition $(\overline{23})$ forms. In that case, all of the imputations, where

$$a \leqslant x_1 < c,\ x_2 > 0,\ x_3 > 0, \qquad (12.19)$$

are stable, together with the coalition structure $\{(\overline{23}), (\overline{1})\}$; which is to say, the object will be sold for any price above and up to but excluding c, and the two buyers will apportion the saving among them in some way.

Indeed, now we have more stable imputations [given the coalition $(\overline{23})$] than there are in all the solutions, because the imputations in which

$$b \leqslant x_1 < c, x_2 > 0, \tag{12.20}$$

which are not in any of the solutions, have been included. R. D. Luce and H. Raiffa, in their analysis of this case, explain it as follows:

"Since players 2 and 3 have agreed not to break up their coalition, the problem first amounts to a bargaining problem between player 1 and coalition $(\overline{23})$, in which it is not obvious that 1 should get less than b. . . ." [28] That is to say, the members of the coalition $(\overline{23})$ will still have something to divide even if the house is sold for more than *b*, and so the seller can hold out for a price larger than *b*. It seems to me, however, that a great deal would depend on the terms of the agreement when the coalition is formed. It seems to me that the only incentive for player 3 to join with player 2 in a coalition is to prevent him from "guaranteeing" the minimum price *b* to the seller.

We turn to the bargaining sets. If the coalition $(\overline{12})$ forms, there is no individually rational payoff configuration which is stable. For, even assuming that player 2 gets nothing and player 1 gets the full value to this coalition, namely *b*, player 1 still has an objection against player 2; for player 1 can get more in a coalition with player 3, whereby player 3 gains. To this, player 2 has no counter-objection.

If, on the other hand, coalition $(\overline{13})$ forms, all the imputations

$$(b \geqslant x \geqslant c, 0, c - x) \tag{12.21}$$

are balanced. Thus we get essentially the core.

In summary, if we ask for "stability," we end up with essentially the imputations in the core. The Von Neumann-Morgenstern solution enlarges the range of possible outcomes but does not guarantee stability. To narrow down the range of outcomes to some particular one of all the possible solutions (N.B. Even a single solution is not a single outcome but a whole range of outcomes.), additional restrictions must be imposed which can be interpreted as "standards of behavior."

13. Large Markets

The conceptualization of a market as a non-zerosum game is quite old. Almost 90 years ago, F. Y. Edgeworth conceived of the elementary market situation as a bargaining problem involving two traders.[29] Initially, each has a quantity of goods of a different kind; say, a blacksmith has nails and a farmer has corn. After a trade each will end up with a combination of the two goods. Assume that the goods are "infinitely divisible"; that is, ignore the impracticality of splitting a single nail or a single grain of corn. Then each trader can end up with any specifiable amount of each commodity, subject only to the restriction that the original total amounts are conserved. (We are assuming no production and no consumption.) Associated with each possible combination of the two commodities is each trader's utility for that combination. We shall assume that each can specify this utility at least on an ordinal scale; that is, each can order all the possible combinations on his scale of preference. This order of preferences is not, of course, necessarily the same for each trader.

Initially, Blacksmith's possession can be designated by (a, 0), which is to say that he has *a* nails and no grains of corn; Farmer's possession is accordingly (0, b): he has no nails and *b* grains of corn.

According to Edgeworth, if the two players are rational, each should have at the end of the trade a distribution which, on his preference scale, is at least as good as his initial distribution. In other words, neither should suffer a utility loss as a result of the trade. Further, no distribution (of the set of all possible distributions) should be preferred by *both* traders to the distribution resulting from the trade (otherwise, being rational, they would effect the preferred one). The set of all distributions which satisfy these two criteria can be represented on a diagram (to be presently described) by a curve called the *contract curve.*

The contract curve is obtained in the following manner. Represent each possible combination of nails and corn in the possession of one of the traders, say of

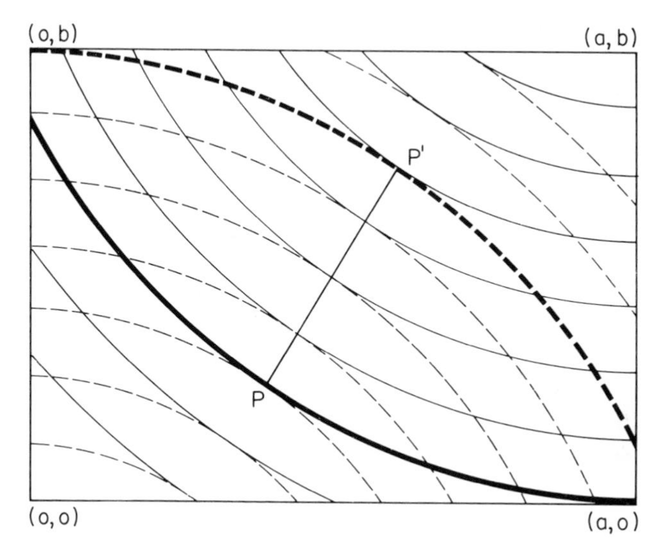

FIG. 16. Solid curves: player 1's indifference curves. Dotted curves: player 2's indifference curves. pp′: contact curve.

Blacksmith, by a pair of numbers (x, y), hence by a point in a two-dimensional space. That is, the horizontal coordinate axis of this space represents nails, the vertical axis corn. It is understood that the remaining nails and corn are in Farmer's possession. Since the total available amounts of both commodities are fixed, the trade space is a rectangle, as shown in Figure 16.

The point (a, 0) represents Blacksmith's initial distribution; the point (0, b) represents what he would have if he traded all of his nails for all of Farmer's corn. The point (a, b) represents the unlikely case where he got all the corn for nothing.

Imagine now the utilities represented in the third dimension; so that the point (x, y) has a certain "altitude" corresponding to the degree of Blacksmith's preference for having that distribution. If Blacksmith's distribution is (x, y), it follows that Farmer's associated distribution is (a − x, b − y), and there is another "altitude" (Farmer's utility) associated with it. Thus there are two "landscapes" over the trade space, one representing Blacksmith's utilities, one Farmer's. On these landscapes we can draw two sets of "contour lines," analogous to contours on topographical maps. Each contour line goes through points of equal "altitude" on one of the landscapes. That is, Blacksmith is indifferent between any two points on one of his contour lines, being willing to trade so many nails for so much corn. These two sets of contours are also shown on Figure 16. As we go in the "northeasterly" direction in the trade space, we assume that we are going "uphill" in Blacksmith's landscape (because he gets more of *both* commodities) and accordingly "downhill" in Farmer's landscape (because he gets less of both).

Edgeworth showed that if the result of the trade is to satisfy the conditions of rationality specified above, then all the possible post-trade distributions must lie on a line joining the points of tangency between pairs of indifference curves of the two traders. In Figure 16 this line is shown as the line PP′.

Note that along this line (the contract curve) the interests of the two traders are in conflict, since going in the general direction toward the point (a, b) gives to Blacksmith more of both commodities and to Farmer correspondingly less. Therefore *where* on the contract curve the final distribution will be depends on how much each trader has his way. *That* the trade will be realized on one of the points of the contract curve is a consequence of the traders' rationality.

Now, this result does not depend on an interpersonal comparison of utilities. (Recall that we assumed only that each trader can order the distributions on his own preference scale.) The contract curve thus represents only the set of outcomes from which the players cannot depart so as to jointly improve their positions. If the utilities could be compared (that is, if it makes sense to speak of the traders' joint utility), a subset could be selected from the set of points which constitute the contract set at which the joint utility is maximized. If this set consists of a single point, it is called the *utilitarian point*. However, if side payments are possible, the singling out of the utilitarian point does not yet constitute a settlement between the two traders. To effect an agreement, it may be necessary for one trader to give a side payment to the other. For example, if the trade is to be treated as a Two-person non-constant-sum game *in characteristic function form*, the solution of this game may be a set of imputations, which implies a range of side payments from one trader to the other, concomitant to the realization of the trade.

Let us now generalize this elementary trading situation to any number of traders. The generalized situation will be called an *Edgeworth game*. The set of players N is a union of two sets: $N = N_1 \cup N_2$. Each member of N_1 is initially in possession of a certain quantity of commodity 1; each member of N_2 of commodity 2. (Imagine n_1 blacksmiths and n_2 farmers coming together to trade.) Each player has a utility function $\psi_i(x, y)$ $(i = 1, 2, \ldots, n)$ which determines the utility (to him) of having in his possession the amount x of commodity 1 and y of commodity 2. We shall assume that

$$\lim_{x\to\infty} (x, y) < \infty \quad \text{and} \quad \lim_{y\to\infty} (x, y) < \infty, \qquad (13.1)$$

i.e., that each commodity, when its quantity becomes very large, brings in diminishing returns of utility (with a "ceiling") to each player.

We now have an N-person game with the following characteristic function:

$$v(\emptyset) = 0 \qquad (13.2)$$

$$v(S) = \max_{x_i, y_i} \left[\sum_{i \in S \cap N_2} \psi_i(a_i - x_i, y_i) + \sum_{j \in S \cap N_2} \psi_j(x_j, b_j - y_j)\right]. \qquad (13.3)$$

This characteristic function can be interpreted as follows. If a set of players form a coalition, they can trade *among themselves* so as to maximize their joint utility. The fact that they are trading among themselves is reflected in the condition $\Sigma x_i = \Sigma x_j$; $\Sigma y_i = \Sigma y_j$. If such a set of traders consists only of members of N_1, for example, then $S \cap N_2 = \emptyset$. and the characteristic function prescribes to the subset of N_1 only $\sum_{i \in S} \psi_i(a_i, 0)$, which is what they would have to begin with, since there is no one in the coalition to trade with.

Depending on the shape of the utility functions, the situation will be represented by different characteristic functions and hence by many different games. To fix ideas, let us select the simplest case where each player's utility function is the same: $\psi(x, y)$. As before, initially each player of N_1 has $(a, 0)$ and each player of N_2 has $(0, b)$. The characteristic function of this game will then be

$$v(\emptyset) = 0 \qquad (13.4)$$

$$v(S) = s\psi\left(\frac{as_{n_1}}{s}, \frac{bs_{n_2}}{s}\right) \qquad (13.5)$$

where $s = |S|$, the number of players in the coalition S; $s_{n_1} = |S \cap N_1|$ $s_{n_2} = |S \cap N|$. This is so, because the identical utility function imposes a perfect symmetry on the game. The situation of each of the players is exactly the same; therefore each player as a member of a trading coalition will end up with the same amount of each commodity. (Note, however, that *how much* he ends up with depends on the size of the coalition he is in.)

We shall now solve the simplest game of this sort à la Von Neumann-Morgenstern. The characteristic function of the Two-person game is

$$v(\emptyset) = 0 \tag{13.6}$$

$$v(1) = \psi(a, 0) \tag{13.7}$$

$$v(2) = \psi(0, b) \tag{13.8}$$

$$v(\overline{12}) = 2\psi(a/2, b/2). \tag{13.9}$$

The Von Neumann-Morgenstern solution singles out the *set* of imputations:

$$\left(2p\psi\left(\frac{a}{2}, \frac{b}{2}\right) - p\psi(0, b) + (1 - p)\psi(a, 0),\right.$$

$$\left.2(1 - p)\psi\left(\frac{a}{2}, \frac{b}{2}\right) + p\psi(0, b) - (1 - p)\psi(a, 0)\right). \tag{13.10}$$

That is to say, none of the imputations in this set dominates any other in the set, and every imputation outside this set is dominated by at least one in the set. There is one imputation for every value of the parameter p.

Let us interpret this result. Note that player 1's share of the joint utility $2\psi(a/2, b/2)$ is

$$2p\psi\left(\frac{a}{2}, \frac{b}{2}\right) - p\psi(0, b) + (1 - p)\psi(a, 0). \tag{13.11}$$

Of this, he originally had $\psi(a, 0)$. Hence his gain is

$$p\left[2\psi\left(\frac{a}{2}, \frac{b}{2}\right) - \psi(0, b) - \psi(a, 0)\right] \tag{13.12}$$

which is a p-th portion of what the two players jointly gained by forming a coalition.

Similarly, player 2's gain is

$$(1 - p)\left[2\psi\left(\frac{a}{2}, \frac{b}{2}\right) - \psi(0, b) - \psi(a, 0)\right]. \tag{13.13}$$

The parameter p remains undetermined by the solution. It represents, in the formulation of Von Neumann and Morgenstern a "standard of behavior." In this case, the parameter can be evaluated as the *market price* of commodity 1 (in relation to commodity 2). The price is "normalized" in the sense that it ranges from 0 to 1. If $p = 0$, commodity 1 is worth nothing on the market, and the utility to player 1 remains the same after the trade as before, namely $\psi(a, 0)$. That is, he ends up with $(a/2, b/2)$ in his possession, but he has paid player 2 an amount of utility equal to $\psi[(a/2, b/2) - (a, 0)]$. If $p = 1$, commodity 2 is worth nothing. Again each player has $(a/2, b/2)$ after the trade, but player 1 has received from player 2 an amount of utility equal to $\psi[(a/2, b/2) - (0, b)]$, namely what player 2 gained in the exchange.

This result is generalizable to any N-person game of this sort, where $N = N_1 + N_2$ (as before) and $N_1 = N_2$; i.e., the number of traders with each of the commodities is equal. That is to say, a set of imputations of the form (13.10), where each member of N_1 receives the first payoff and each member of N_2 the second, is *one* solution. However, if $N_1 > 1$, there are other solutions.

One of the results obtained by Edgeworth was that, if the number of traders increases, the contract curve shrinks, approaching a single point as the number of traders becomes infinitely large. This point (being in n-dimensional space) represents the distribution of the goods among the n players (n being very large) under conditions of free competition. The basic idea is that when additional traders are available, and when the traders are free to seek better bargains, an equilibrium will be eventually established. Let us see how this situation is reflected in the N-person game.

From the form of the solution (13.10) we see that it does not shrink to a single imputation as n increases, since p does not depend on n and preserves its range from 0 to 1. What does happen, however, is that the *core* of the game shrinks to a single point as the number of players increases without bound. Recall that the core is a subset of the set of imputations which are not dominated (via any coalition) by any other imputation. Let us, therefore, shift our attention to the core.

First, let us examine the market with a monopolist. That is to say, a single trader initially possesses commodity 1; all others (n in number) possess only commodity 2. Thus $n_1 = 1$; $n_2 = n$. Then the following imputation lies in the core:

$$\eta = \left((n+1)\psi\left(\frac{a}{n+1}, \frac{nb}{n+1}\right) - n\psi(0, b), \right.$$
$$\left. \psi(0, b), \psi(0, b) \ldots \psi(0, b)\right). \quad (13.14)$$

Recall that $\psi(0, b)$ was the utility with which the members of N_2 started. As a result of trade, each ends up with an equal share of each of the commodities. The utility to the grand coalition is $\left[(n+1)\psi\left(\frac{a}{n+1}, \frac{nb}{n+1}\right)\right]$. If imputation (13.14) obtains, the monopolist has gotten all of the gain which has accrued to the coalition from the trading; that is, he keeps the total utility after having paid off to the members of N_2 the utilities with which they started.

To show that imputation (13.14) is in the core, it is sufficient to show that no subset of the $n + 1$ players, acting as a coalition, can receive jointly more than they get in η.

To be sure, (13.14) is not the only imputation in the core. There are others in which the monopolist gets less; say, a quantity ε less (and accordingly, the others get more). However, it can be shown that, as the number of members of N_2 increases, the range of this ε shrinks; that is, any imputation η' in which the monopolist gets less than $(n+1)\psi\left(\frac{a}{n+1}, \frac{nb}{n+1}\right) - \varepsilon$ lies outside the core if the number of members of N_2 is sufficiently large. It follows that, in a market with one monopolist and an infinitely large number of traders with a commodity which they want to trade for the monopolist's commodity, the entire utility gain accruing to *everyone* from the trade is appropriated by the monopolist *if* the outcome is to be in the core; i.e., if the outcome is to be an imputation dominated by no other via any coalition of a subset of players.

It is easy to see intuitively the reason for this result. The members of N_2 all compete with each other to enter a coalition (i.e., a trading agreement) with the monopolist. The monopolist does not compete with anybody, but his "cooperation" is essential for any profitable transaction. In practice, the monopolist may not be able to get away with paying a merest pittance. However, if there is an inexhaustible market for his commodity, he can say to anyone who wants to trade "Take this pittance or leave it; there is always another who will take less."

If we admit solutions which are *not* in the core, this amounts to allowing coalitions which *are* dominated (via some subsets) by other coalitions. This is interpreted to mean that some subset of traders made up their minds to "stick together" and not to accept anything less than some minimum utility bonus. It may be of advantage to some of the members of this coalition to "leave it" (i.e., to make a separate deal with the monopolist) whereby they would get a bit more and leave nothing to their erstwhile allies. If they do not yield to this "temptation," that is, preserve the original coalition, the monopolist will not be able to get the entire gain. This situation has already been discussed in the case of the small market with the one seller and two buyers. (The seller was the "monopolist." His commodity was his house, the only house on the market. The buyers' commodity was money.)

Let us now look at a large market with several traders in each of the two commodities. Let $n_1 = kn_1'$ and $n_2 = kn_2'$. Eventually we shall assume that k becomes very large, in other words, that the numbers of traders in each set N_1 and N_2 increases proportionately. The characteristic function, as has been said, is

$$v(\emptyset) = 0 \tag{13.15}$$

$$v(S) = s\psi\left(\frac{s_{n_1}a}{s}, \frac{s_{n_2}b}{s}\right). \tag{13.16}$$

If we hold $n = n_1 + n_2$ fixed, the joint utility to coalition S consisting of s_{n_1} players from N_1 and s_{n_2} from N_2 will vary depending on the ratio $n_1 : n_2$. This joint utility will have a maximum value for some particular value of this ratio. It can be shown that this ratio will maximize the joint utility to any of all possible coalitions. It thus represents the "optimal" ratio of the two classes of traders. (Think of this ratio as being determined by the sort of commodities being traded.) In what follows, we assume that $n_1 : n_2$ is in fact this optimal ratio.

Now we can assert that the imputation

$$\eta \equiv \left(\frac{v(N)}{n}, \frac{v(N)}{n}, \frac{v(N)}{n}, \ldots, \frac{v(N)}{n}\right) \tag{13.17}$$

is always in the core (for all values of k, hence of n).

The imputation η is, of course, the "egalitarian" imputation in which every member of N receives the same utility. As before, there are other imputations in the core, which can be obtained from η by redistributing the payoffs so that some get more and some less—but only within limits set by the magnitude of n (hence of k). The principal result is that as k (hence n) becomes very large, these limits tend to become ever narrower; so that in the limit, the egalitarian imputation is the only one remaining in the core.

This result holds only if the "optimal" ratio of the two types of traders is maintained. If it is not, then as the number of players increases, a subset of players can always be found who *can* obtain the jointly maximal return. The imputation which reflects this will dominate any imputation and will itself be dominated by another imputation via some subset of players. In other words, there will be no undominated imputation, and the core will be empty.

Next, we shall examine a "total economy" as an N-person game in characteristic function form. We shall say that an economy has *increasing returns* if it is true that whenever $T \supset S$, then $\frac{v(T)}{T} > \frac{v(S)}{S}$.

In words, increasing the size of a coalition makes possible an increase in the utility *of each individual member* of the coalition. In such an economy, additional members of a coalition are always "welcome," and in the grand coalition it is possible to achieve a distribution of payoffs such that each member will receive more than he could receive in any other coalition. Such a game will always have a core (for any number of players). In particular, the egalitarian imputation

$$\vec{e} = \left(\frac{v(N)}{n}, \frac{v(N)}{n}, \ldots, \frac{v(N)}{n}\right) \qquad (13.18)$$

will always be in the core. (This can be easily seen, since no smaller subset in a coalition by themselves can jointly get more than they can get in e, and so the only effective set for e is N.)

An economy is said to have *decreasing returns* if for all subsets $S, T \subset N$, it is true that whenever $S \supset T$, $\frac{v(S)}{S} < \frac{v(T)}{T}$. In such an economy additional members are not welcome in any coalition. (They are "extra mouths to feed," who do not contribute enough to make it worthwhile.) For such a game no core exists, and the grand coalition is "unstable." In this way some reasonable conjectures about the relation between the nature of an economy and the conflicts which are likely to arise in it are consequences of formal game-theoretical analysis.[30]

14. Simple Games and Legislatures

We have seen that the characteristic function description of a game amounts to specifying a function v( ) whose domain varies over all the subsets of N players and whose range is some subset of the real numbers. If the range of this function is exactly two numbers, the game is said to be *simple.* In other words, in a simple game some of the coalitions command one value and the remaining coalitions command another. Without loss of generality we can designate the smaller value by 0 and the larger value by 1. Then all the coalitions which get 1 are called *winning* coalitions. In addition, the definition of a simple game comprises the following criteria.

1. Any coalition which includes a winning coalition is itself a winning coalition.
2. Some coalitions are winning coalitions.
3. The coalition without members (i.e., one whose set of players is empty) is not a winning coalition.

The first criterion is in accordance with most observations that the "power" of a coalition cannot decrease if additional players join it. The criterion preserves the super-additive property of the characteristic function. The second is assumed to exclude the trivial case of games without winning coalitions. The third is assumed in order to exclude the equally trivial case in which every coalition is a winning coalition. (If the empty set were a winning coalition, every set would also be one, because every subset includes the empty set.)

The coalitions of a simple game which are not winning coalitions are called *losing coalitions.* Thus, if $\mathfrak{N}$ is the set of all the subsets of N, $\mathcal{W}$ the set of winning coalitions, and $\mathcal{L}$ the set of all losing coalitions, then

$$\mathcal{L} = \mathcal{N} - \mathcal{W}. \tag{14.1}$$

We shall need the following notation. (Recall that we use script capital letters to denote sets of sets [cf. p. 22].)

$\mathcal{S}$ ≡ any set of sets
$\mathcal{S}^+$ ≡ the set of supersets of $\mathcal{S}$.
$\mathcal{S}^-$ ≡ the set of subsets of $\mathcal{S}$.
$\mathcal{S}^*$ ≡ the set of complements of $\mathcal{S}$.
$\bigcap \mathcal{S}$ ≡ the intersection of the elements (which are sets) of $\mathcal{S}$.
$\bigcup \mathcal{S}$ ≡ the union of the elements of $\mathcal{S}$.

Note that $\mathcal{S}^*$ is *not* the complement of $\mathcal{S}$ with respect to some set of sets $\mathcal{N}$. For example, if $\mathcal{N}$ is the set of all sets of three elements, then

$$\mathcal{N} = \{(1), (2), (3), (1, 2), (1, 3), (2, 3), (1, 2, 3), \emptyset\}. \tag{14.2}$$

Suppose $\mathcal{S} = \{(1), (1, 2), (2, 3), \emptyset\}$.
Then the complement of $\mathcal{S}$ with respect to $\mathcal{N}$ is

$$\mathcal{N} - \mathcal{S} = \{(2), (3), (1, 3), (1, 2, 3)\}. \tag{14.3}$$

But $\mathcal{S}^* = \{(2, 3), (3), (1), (1, 2, 3)\}$. (14.4)

We have already defined the sets of sets $\mathcal{N}$, $\mathcal{W}$, and $\mathcal{L}$ for an N-person game.

The set of sets $\mathcal{B} \equiv \mathcal{L} \cap \mathcal{L}^*$ is the set of the *blocking* coalitions of the game.

In other words, the losing coalitions, as they are defined here, are the coalitions which are not winning coalitions. Since $\mathcal{L} \cap \mathcal{L}^* \subset \mathcal{L}$, it follows from the definition of blocking coalitions that they are all included in the losing coalitions. In other words, some of the losing coalitions may be blocking coalitions. (N.B. A game may have no blocking coalitions. This happens, of course, if $\mathcal{L} \cap \mathcal{L}^* = \emptyset$.)

Before we go further, let us illustrate the types of coalitions by an example of a simple game.

Let there be four players: 1, 2, 3, 4. Then $\mathcal{N}$ contains 16 sets:

$$\mathcal{N} = \{(1), (2), (3), (4), (1, 2), (1, 3), (1, 4), (2, 3), (2, 4), (3, 4), (1, 2, 3), (1, 2, 4), (1, 3, 4), (2, 3, 4), (1, 2, 3, 4), \emptyset\}. \quad (14.5)$$

Let us define as the winning coalitions all the subsets of N (i.e., sets of members of $\mathcal{N}$) with at least three members (players). These constitute

$$\mathcal{W} = \{(1, 2, 3, 4), (1, 2, 3), (1, 2, 4), (1, 3, 4), (2, 3, 4)\}. \quad (14.6)$$

Therefore the remaining coalitions are losing coalitions. These comprise the set of sets

$$\mathcal{L} = \{(1), (2), (3), (4), (1, 2), (1, 3), (1, 4), (2, 3), (2, 4), (3, 4), \emptyset\}. \quad (14.7)$$

The set (of sets) $\mathcal{L}^*$ contains the complements of the set $\mathcal{L}$, which comprises $\mathcal{W}$ and also the pairs

$$(1, 2), (1, 3), (1, 4), (2, 3), (2, 4), (3, 4). \quad (14.8)$$

These pairs are in both $\mathcal{L}$ and $\mathcal{L}^*$, since each of the pairs is a complement of another pair.

It follows that $\mathcal{B}$ is the set of all pairs.

In this connection, it is well to note that while the complement of $\mathcal{L}$ in $\mathcal{N}$ is $\mathcal{W}$ (i.e., the *set* of coalitions which are not losing coalitions is the set of winning coalitions), it is not true that the complement of a particular losing coalition L is necessarily a winning coalition. Thus in our example, both (1, 2) and (3, 4) are losing coalitions.

Now let us interpret the result. Imagine a committee of four members which can take action only if at least three members approve. Then the entire committee and all the triples are winning coalitions (in the sense that they can get the committee to act). All the other subsets are losing coalitions (in the sense that they cannot get the committee to act). But all the pairs are also blocking coalitions in the sense that they can *prevent* the committee from acting (although they cannot get the committee to act).

Note that if a committee had an odd number of members, each with one vote, and took action by majority vote, there would be no blocking coalitions in it. The reader can verify this result formally by enumerating all the subsets of $\mathcal{W}$, $\mathcal{L}$, and $\mathcal{L}^*$. Here the crucial circumstance is that $\mathcal{L} \cap \mathcal{L}^* = \emptyset$; hence $\mathcal{B}$ is empty.

Simple games without blocking coalitions are called *strong* games. We can see readily that if n, the number of players, is odd, and if every majority of players is a winning coalition, the game is strong. At the other extreme, consider a game in which the only winning coalition is the grand coalition N. For example, if the players are members of a committee in which unanimous consent is needed to render a decision, then every player by himself constitutes a blocking coalition. The American jury is the best known example of a Twelve-person simple game in which every player is a blocking coalition, and consequently, the grand coalition is the only winning coalition. Such games are called *pure bargaining games.* More generally, games in which at least one player is a member of every winning coalition are called *weak.* That is, in a weak game $\bigcap \mathcal{W} \neq \emptyset$. In a weak game, at least one player (but not necessarily every player) has a *veto.*[31]

For completeness, we shall also mention the *null* game, in which there are no winning coalitions. (Note that the null game does not satisfy the criteria of a simple game.) Null games are in themselves totally uninteresting; but they are convenient as constructs in the definition of other games, as we shall see.

If the only criterion which makes a subset of players a winning coalition is the number of players in it, we shall call the corresponding game a *simple legislature.* Here n designates the number of players, and k the number of players needed to form winning coalitions. Specifically, if n is odd and any simple majority forms a winning coalition, we have the game designated by $M_{n,\frac{1}{2}(n+1)}$. If n is even and it takes a simple majority to win, we have $M_{n,\frac{1}{2}(n+2)}$. The pure bargaining game corresponds to $M_{n,n}$. The null game can be symbolized by $M_{n,n+1}$.

In some real legislatures, committees, etc., it takes more than a majority to pass certain measures. Thus $M_{25,17}$ could designate a body of 25 members which can act only with consent of at least two-thirds of the members.

Consider now a game $M_{n,k}$, where $k < n/2$. Here a minority can be a winning coalition. In practice, this might make for difficulties, since in this case more than one winning coalition could form and, without further information, we could not tell to which of these the payoff of 1 (e.g., the "power to act") accrues. Such games are called *improper*. In an improper game $\mathcal{W} \cap \mathcal{W}^* \neq \emptyset$. Like null games, improper games are useful constructs in the analysis of certain more complex games.

L. S. Shapley[32] gives as an example of an improper game the alternative methods of amending the U.S. Constitution, namely by two-thirds majority in both Houses of Congress (with ratification by three-fourths of the state legislatures), and by three-fourths of special state conventions with or without the approval of Congress. However, this example seems to me to be erroneous, since the state legislatures are involved in both methods, either in ratifying the proposed amendment (if proposed by Congress) or in initiating the amendment (if it is to be ratified by state conventions). Thus the winning coalitions are not mutually exclusive sets. A better example of an improper game is a situation where a determined, organized minority can seize dictatorial power. If two such groups were formed, before any of them could act we could not decide "by the rules of the game" which group was the "winning coalition." Presumably the issue would be decided by other factors than the above-mentioned rule.

Note: By-laws of legislative bodies, election procedures, etc., often empower minorities to "have their way," as, for example, requiring a party to be put on the ballot if, say five percent of the electorate so petition. This situation does not constitute an improper game in the sense defined, since there is no provision for another minority coalition to prevent the party from getting on the ballot. A legislature (committee, electorate) would be an improper game only if its rules allowed two or more mutually exclusive sets of members to have a decisive voice on the *same issue*.

We now define the set of *minimal* winning coalitions as

$$\mathcal{W}^m \equiv \bigcap \text{ of all } \mathcal{S} \text{ such that } \mathcal{S}^+ = \mathcal{W}. \qquad (14.9)$$

In other words, take all sets of sets $\mathcal{S}$ such that the sets composed of the supersets of members of $\mathcal{S}$ all comprise the set of winning coalitions. Then $\mathcal{W}^m$ is the intersection of all such $\mathcal{S}$. Thus, in a game $M_{n,k}$, all sets of exactly k players constitute $\mathcal{W}^m$, the set of minimal winning coalitions.

A player who is not in any minimal winning coalition is called a *dummy* player. To put it another way, any winning coalition can expel a dummy player without thereby becoming a losing coalition. On the other hand, a player who is a member *only* of winning coalitions is called a *dictator*. For it follows from the definition that the coalition consisting of this player alone is a winning coalition; and moreover, no coalition can be a winning one without him. Hence, if the stake of the game is the power to act, the dictator alone has this power. Note that if there is more than one dictator, the game is improper. Note also that games with dictators are inessential games (cf. p. 83), since there is no advantage for anyone to join a coalition. To put it another way, the characteristic function of a proper game containing a dictator is simply additive, never strictly super-additive, since the union of two losing coalitions is always a losing coalition.

A simple game is completely defined (in its characteristic function representation) when the number of players and the set of winning coalitions have been specified. Thus a simple game can be specified by a pair

$$G = (N, \mathcal{W}). \tag{14.10}$$

The game *dual* to G is

$$G^* = (N, \mathcal{L}^*). \tag{14.11}$$

That is to say, the players of a dual game are the same as those of the original game, while the winning coalitions are the complement sets of the losing coalitions of the original game. It follows that if a game has no blocking coalitions (i.e., is strong), the dual game is identical with it; i.e., $G^* = G$. This is because the complement of any losing coalition of a strong game G is a winning coalition of G; hence the sets of winning coalitions are identical in the two games. If, however, the game G has blocking coalitions, then the winning coalitions of the dual game include (besides the winning coalitions of G) also the blocking coalitions of G. This is because the complement of a blocking coalition is also a blocking coalition (recall that the blocking coalitions are also among the losing coalitions). In terms of a legislative body, if in the original game "to win" means to pass a measure, in the dual game "to win" means to prevent a measure from passing. Thus all the blocking coalitions of the original game become winning coalitions in the dual game. The winning coalitions, of course, remain winning coalitions, because any coalition (in a game which is not improper) able to pass a measure can also prevent a measure from being passed.

Games can be viewed as "mathematical objects"; therefore operations can be performed on them as on any other mathematical objects, if these operations are properly defined. We shall define two such operations, namely the *product* and the *sum* of two games. Given two simple games $(N_1, \mathcal{W}_1)$ and $(N_2, \mathcal{W}_2)$, where $N_1 \cap N_2 = \emptyset$ and $N_1 \cup N_2 = N$, we write the product as

$$(N, \mathcal{W}) = (N_1, \mathcal{W}_1) \times (N_2, \mathcal{W}_2). \qquad (14.12)$$

Since we have already defined N, it remains to define $\mathcal{W}$, in order to specify the product game completely. In symbols,

$$\mathcal{W} = \{S \mid S \cap N_1 \in \mathcal{W}_1 \textit{ and } S \cap N_2 \in \mathcal{W}_2\}. \qquad (14.13)$$

In words, the winning coalitions of $(N_1, \mathcal{W}_1) \times (N_2, \mathcal{W}_2)$ are those subsets of N whose members are in winning coalitions in $(N_1, \mathcal{W}_1)$ if they belong to N_1, and in winning coalitions of $(N_2, \mathcal{W}_2)$ if they belong to N_2.

The sum of two games is defined analogously. If

$$(N, \mathcal{W}) = (N_1, \mathcal{W}_1) + (N_2, \mathcal{W}_2), \qquad (14.14)$$

then

$$N = N_1 \cup N_2, \mathcal{W} = \{S \mid S \cap N_1 \in \mathcal{W}_1 \textit{ or } S \cap N_2 \in \mathcal{W}_2\}. \qquad (14.15)$$

In other words, those players who were in winning coalitions *either* in $(N_1, \mathcal{W}_1)$ *or* in $(N_2, \mathcal{W}_2)$ are members of winning coalitions of $(N, \mathcal{W})$.

The operations of "addition" and "multiplication" are both commutative and associative. That is

$$(N_1, \mathcal{W}_1) \times (N_2, \mathcal{W}_2) = (N_2, \mathcal{W}_2) \times (N_1, \mathcal{W}_1); \qquad (14.16)$$

$$[(N_1, \mathcal{W}_1) \times (N_2, \mathcal{W}_2)] \times (N_3, \mathcal{W}_3) = (N_1, \mathcal{W}_1) \times [(N_2, \mathcal{W}_2) \times (N_3, \mathcal{W}_3)]; \qquad (14.17)$$

$$(N_1, \mathcal{W}_1) + (N_2, \mathcal{W}_2) = (N_2, \mathcal{W}_2) + (N_1, \mathcal{W}_1); \qquad (14.18)$$

$$[(N_1, \mathcal{W}_1) + (N_2, \mathcal{W}_2)] + (N_3, \mathcal{W}_3) = (N_1, \mathcal{W}_1) + [(N_2, \mathcal{W}_2) + (N_3, \mathcal{W}_3)]. \quad (14.19)$$

One might almost think that a distributive law would be in force (as in the analogous operations on sets). However, the game $G_1 \times (G_2 + G_3)$ is not necessarily identical with the game $(G_1 \times G_2) + (G_1 \times G_3)$. We do have the following relations, which are analogous to those governing the operations of union, intersection, and complementation of sets (cf. p. 20):

$$(G + H)^* = G^* \times H^* \quad (14.20)$$

$$(G \times H)^* = G^* + H^*. \quad (14.21)$$

Let us now examine two real life examples of product games. Consider the U.S. Congress. Define "winning" as passing a bill, which requires a majority of both houses. The House of Representatives is a majority game with 437 players. According to our designations of this game, we shall denote it by $M_{437,219}$. Analogously the Senate is denoted by $M_{101,51}$ (we must count the Vice-President, who can break ties). Congress as a whole is, therefore, a product game, $M_{437,219} \times M_{101,51}$.

Note that Congress as a whole is not a majority game, since not every coalition comprising a majority of Congress as a whole (as distinguished from majorities in both houses) is a winning coalition. Congress is not a strong game because, among all the subsets of its members, many can form blocking coalitions. It can be easily shown that a product game can never be strong.

Consider next the United Nations Security Council. The Big Five comprise a pure bargaining game, because each member has a veto. Define a winning coalition among the non-permanent members to be any coalition consisting of at least four members. Clearly, then, the game in which the non-permanent members of the Security Council are the players is, as defined, an improper game (because more than one "winning coalition" can form simultaneously). Nevertheless the product game, which is the entire Security Council, is proper. A winning coalition is defined as any coalition consisting of the Big Five and any four or more non-permanent members. At any one time there can be at most one such winning coalition.

It is also generally true that if at least one of the components of a product game is proper, the product also is proper. Likewise it can be shown that a product is weak if and only if at least one of the component games is weak. (Clearly if no player has a veto in either component, then no player has a veto in the product; conversely, if a player has a veto in a component, he has a veto in the product.) A weak legislature (such as the United Nations Security Council) can always be factored into two components, one of which is a pure bargaining game.

Since legislatures which are sums of legislatures are improper, these are not common in real life. However, product games might well exist in which one component is a sum of legislatures. Consider a system of lobbies and Congress as a legislative unit. Imagine that the initiatives for legislative acts come from the lobbies but that legislation is enacted only with the approval of Congress. Then the lobbies by themselves may be viewed as an improper legislature (where "winning" is defined by success in initiating a piece of legislation). They can also be viewed as a sum of legislatures (each represented by a lobby), since a piece of legislation is initiated if it originates in *any* of the lobbies. The product of the lobbies and Congress is a proper game.

The notions of sum and product suggest a more general notion, that of a compound game. An elementary example of such a game is the U.S. presidential election. Let us assume for simplicity that all the electors chosen by the people of each state must be committed to one of the presidential candidates. (We are excluding situations such as arose in the elections of 1964 when the electors of some states split.) Then the elections in the several states are 53 games (the District of Columbia, Puerto Rico, and the Virgin Islands included), namely the elections of the electors by the voters. The outcomes of these 53 games determine a "supergame," namely the election of the president by the electors. It turns out that this supergame is of the type known as a *weighted majority game,* to be discussed below.

In an N-person weighted majority game, each of the n players may have more than one "vote." The winning coalitions are those which together have some requisite number of votes. The election of the president of the United States by the electoral college (assuming that the electors of every state vote as a bloc) has already been mentioned. Other well known examples are stockholders' meetings of corporations, where individual stockholders have voting power in proportion to the number of shares they own. A weighted majority game can be represented by the symbol

$$(q; w_1, w_2, \ldots, w_n) \qquad (14.22)$$

where the w_i are the weights (i.e., the number of votes possessed by player i) and q is called the *quota,* that is, the number of votes required to win.

Note that weights can be assigned in more than one way without changing the set of winning coalitions. Since the simple game is determined completely when the set of players N and the set of winning coalitions $\mathcal{W}$ have been specified, it follows that the weights can be assigned in many ways without changing the game itself in any essential manner. For instance, consider a Four-person game in which the winning coalitions are $(\overline{12})$ (minimal), $(\overline{123})$, and $(\overline{1234})$. This game can be represented by (3; 3/2, 3/2, 1/2, 1/2) and equally well by (4; 2, 3, 1/2, 1/4), or by any number of such designations. We can therefore choose q and w_i at our convenience. In particular, they can always be chosen to be integers.

If we can choose the weights so that all the minimal winning coalitions have the same combined weight, the game is called *homogeneous;* otherwise inhomogeneous. It can be shown that all Three-person weighted majority games and all proper Four-person weighted majority games are homogeneous. However, examples of Five-person inhomogeneous games exist, e.g.,

$$(5; 2, 2, 2, 1, 1). \tag{14.23}$$

Here the minimal winning coalition $(\overline{123})$ has combined weight of 6, while the minimal winning coalition $(\overline{124})$ has combined weight of 5. The weights cannot be reassigned to give all minimal winning coalitions the same combined weight.

Note that this game is not strong since it has blocking coalitions. All Five-person inhomogeneous games can be shown to be not strong.[33] However, strong inhomogeneous weighted majority games with more than five players exist. For example,

$$(5; 2, 2, 2, 1, 1, 1). \tag{14.24}$$

Some simple games can be represented both as products and as weighted majority games. For example, the United Nations Security Council can be represented as $B_5 \times M_{10,4}$ where B_5 is a pure bargaining game (the Big Five being the players) and $M_{10,4}$ is an improper game. It can also be represented as a weighted majority game

$$(39; 7, 7, 7, 7, 7, 1, 1, 1, 1, 1, 1, 1, 1, 1, 1). \tag{14.25}$$

That is to say, if the voting rules of the Security Council were such that 39 votes were required to render a decision, and each of the Big Five had seven votes while each of the non-permanent members had one vote, then the set of winning coalitions would be exactly the same as under the present rules (the Big Five and at least four non-permanent members).

Note that this game is both homogeneous and dual homogeneous. (That is, the dual of this game, wherein the winning coalitions are the original ones plus the blocking coalitions, is also homogeneous.)

If a game is both homogeneous and dual homogeneous, its weighted majority representation gives a good idea of the relative "power" of the players.

Not all simple games can be represented as weighted majority games.

An example is $M_3 \times M_3$, which might represent a two-chamber legislature of three members each, with a simple majority in both houses constituting a minimal winning coalition. If the weights of the winning coalitions could all be made equal, and if there were some number that could serve as a quota, then (since the players are not distinguishable) specifying the number of players in a coalition would determine whether it was a winning coalition or not. But in this game, some sets of four players are winning (two in each component) and some are losing (all other such sets). Therefore this game cannot have a quota, and so cannot be represented as a weighted majority game.

Questions which naturally arise with regard to simple games concern the relation between the solutions offered by game theory and our intuitive notions about what the outcomes of these games ought to be. In Chapter 19 we shall examine some of these notions brought out by behavioral scientists. One is the so-called principle of minimum resources, according to which the coalition expected to form would be a coalition whose members jointly possess the minimal amount of "resources" (in this case weights or "votes") to win the prize of the game. The other is the so-called principle of minimum power, according to which that coalition is expected to form whose members jointly possess the minimal amount of "power" (as calculated by means of the Shapley value).

In addition to these expectations, behavioral scientists also offer some thoughts on how the winning coalition distributes the prize among its members. In connection with the minimum resource theory, at least, the hypothesis has been proposed to the effect that the prize shall be distributed in proportion to the "resources" which the respective members have brought into the coalition (e.g., the number of votes the individuals possess).

As we shall see in Chapter 19, although experimental evidence indicates that the members of a winning coalition with more votes do tend to receive larger shares of the prize, the shares tend to be less than proportional to the "weights" of the players.

From the game-theoretic point of view, it is clear that within a wide range of weight distributions, the weights play no part in the strategic considerations of coalition formation. That is, once the winning and blocking coalitions of a simple game have been specified, any redistribution of weights which does not change the sets of winning and blocking coalitions has no significance for the bargaining positions of the players except, perhaps, in the subjective estimations of the players, with which game theory is not concerned. (The players are supposed to be rational and so are supposed to see what a player can actually *do* in a game, regardless of what his capabilities *appear* to be.) Therefore, within a wide range of "resource" distributions, the "resources" of the players are irrelevant to the game-theoretic solutions of such games.

The minimal winning coalitions, however, do play a part in the theory. For example, if we demand that a Von Neumann-Morgenstern solution be *symmetric* (that is to say, all the imputations in it are obtainable from each other by a permutation of the players) then, as has been shown by Raoul Bott,[34] the only such solution of the simple majority game $M_{n,k}$ where n is odd and $k = \frac{1}{2}(n + 1)$ is the set of imputations in which any minimal winning coalition divides the prize equally. However, if more than a simple majority is required to win, there is an infinity of symmetric solutions. Moreover, if the "symmetric standard of behavior" is dropped, it is not true that, even in the simple majority game, every solution awards positive payoffs only to a minimal majority. We have seen, for example, that in the Three-person constant-sum game, any set of imputations

$$\vec{x} = (c, x_2, x_3) \tag{14.26}$$

where c is a constant $(0 \leqslant c < \frac{1}{2})$, $x_2 + x_3 + c = 1$, is a solution. Most imputations of a solution of this sort in which $c > 0$ award positive payoffs to every player, not just to a minimal majority. Clearly, such a solution is not symmetric. The only symmetric solution of this game is the set of imputations

$$(0, \tfrac{1}{2}, \tfrac{1}{2}), (\tfrac{1}{2}, 0, \tfrac{1}{2}), (\tfrac{1}{2}, \tfrac{1}{2}, 0), \tag{14.27}$$

which *is* a minimal majority solution.

It appears, then, that the intuitively acceptable notion that, in a simple majority game, a minimal winning coalition will split the prize equally, is substantiated by the Von Neumann-Morgenstern theory if an additional assumption (a symmetric standard of behavior) is made.

We shall return to some applications of the theory of simple games and legislatures in Chapter 19.

15. Symmetric and Quota Games

Symmetric Games

Symmetric games are those in which the value of a coalition depends only on the number of players in it. In other words, the strategic position of each player in the process of coalition formation is the same. Clearly, then, the Shapley value and the Harsanyi bargaining model award the same payoff to every player.

A simple legislature $M_{n,m}$ (cf. Chapter 14) is a special case of a symmetric game, whose characteristic function is

$$v(S) = 0, \text{ if } |S| < m, \quad (15.1)$$

$$v(S) = 1, \text{ if } |S| \geqslant m. \quad (15.2)$$

In a symmetric game, this characteristic function is generalized so that it reads

$$v(S) = v(q), \text{ if } |S| = q \quad (q = 0, 1, 2, \ldots, n). \quad (15.3)$$

In what follows we shall write v_q for $v(q)$, the value of the game to any coalition with q players. We shall also assume that $v(S)$ is super-additive.

It is useful, in this context, to introduce the idea of the *pro-rata gain,* that is, a player's share in the value of the coalition to which he belongs, assuming that this payoff is equally divided among the members of the coalition. Thus the pro-rata gain of a coalition with q members is v_q/q. Since v_q/q takes only a finite number of values (at most $n + 1$), some one of them (or several) must be the largest among them. Accordingly, we define

$$a = \operatorname*{Max}_{q} (v_q/q), \quad (15.4)$$

that is, the largest pro-rata gain which can be achieved by any coalition. If $v_q/q = a$ for several values of q, one of these q's must be the smallest. Accordingly, we define p as the smallest of the q's for which $v_q/q = a$; that is

$$p = \operatorname*{Min}_{q} [q : v_q/q = a]. \qquad (15.5)$$

Clearly, $p \leqslant n$.

Finally, if upon dividing n by p we obtain a remainder, we designate it by r, so that

$$n = mp + r \quad (0 \leqslant r < p). \qquad (15.6)$$

In what follows, we shall examine zerosum symmetric games. We normalize our game so that

$$v(\bar{i}) = -1; v(N) = 0 \quad (i = 1, 2, \ldots, n). \qquad (15.7)$$

Then v_{n-1} must equal 1. Thus $v_{n-1}/(n-1) = 1/(n-1) > 0$, and a, being the largest of the pro-rata gains must likewise be positive: $a > 0$.

Next, observe that n cannot be divisible by p; i.e., $r \neq 0$. For if it is, we have

$$n = pm, 0 = v_n \geqslant mv_p = mpa, \qquad (15.8)$$

so that $a \leqslant 0$, which contradicts $a > 0$.

If q, the number of players in a coalition, is divisible by p, then $v_q = qa$; that is, each player in this coalition receives the maximum pro-rata gain. This is shown as follows:

If $q = kp$ (k being an integer), then by the super-additive property $v_q \geqslant kv_p$. But then $v_q \geqslant kv_q = kpa = qa$, so that $v_q \geqslant qa$. But since a is the maximal pro-rata gain, we must have $v_q \leqslant qa$. Therefore $v_q = qa$.

A coalition S is said to be *flat* if its value is the sum of the values to the players that compose it. Thus, in our symmetric game, S is flat if $v(S) = -q$, where $|S| = q$. If S with q players is flat, then $q \leqslant r$. For if $q > r$, then $v_r = -r$ and $0 < q = v_{n-q} > r = v_{n-r}$. Then

$$\frac{v_{n-q}}{n-q} > \frac{v_{n-r}}{n-q} > \frac{v_{n-r}}{n-r} = \frac{v_{mp}}{mp} = \frac{mpa}{mp} = a. \quad (15.9)$$

But by the definition of a, we must have

$$a \geqslant \frac{v_{n-q}}{n-q}. \quad (15.10)$$

Thus $q > r$ leads to a contradiction.

Finally, we must have the following inequalities satisfied:

$$\frac{1}{n-1} \leqslant a \leqslant \frac{r}{pm} < \frac{1}{m}. \quad (15.11)$$

For $1/(n-1) = v_{n-1}/(n-1) \leqslant a$; $v_{pm} = pma = -v_r \leqslant r$. Also $r < p$ [cf. (15:6)].

Example

Let us check all of these findings in a particular symmetric zerosum game. Let $n = 12$, and let the characteristic function be

$$v_1 = -1;\ v_2 = -2;\ v_3 = -9/5;\ v_4 = -8/5;$$
$$v_5 = -1;\ v_6 = 0;\ v_7 = 1;\ v_8 = 8/5;\ v_9 = 9/5;$$
$$v_{10} = 2;\ v_{11} = 1;\ v_{12} = 0. \quad (15.12)$$

Then the pro-rata gains are given by

$$g_1 = -1;\ g_2 = -1;\ g_3 = -3/5;\ g_4 = -2/5;$$
$$g_5 = -1/5;\ g_6 = 0;\ g_7 = 1/7;\ g_8 = 1/5;\ g_9 = 1/5;$$
$$g_{10} = 1/5;\ g_{11} = 1/11;\ g_{12} = 0. \quad (15.13)$$

The largest pro-rata gain is 1/5. The smallest coalition which awards the largest pro-rata gain to its members is a coalition of 8 players. Upon dividing 12 by 8, we get a remainder of 4. Therefore in this case

$$a = 1/5;\ p = 8;\ m = 1;\ r = 4. \quad (15.14)$$

We see that $a > 0$ and $r > 0$, as required. Since $p = 8$, there is no $q \leqslant n$ which is divisible by p; hence the result proved above ($v_q = qa$) does not apply. The only flat coalitions are those with two members, and $2 < 4$, as required. Finally, $1/(n-1) = 1/11$, $a = 1/5$, $\frac{r}{pm} = 1/2$, $1/m = 1$, so that inequalities (15.11) are satisfied.

In applying the Von Neumann-Morgenstern theory of solutions, game theoreticians are usually interested in singling out special solutions which can be somehow intuitively interpreted. For example, in a symmetric game, symmetric solutions are clearly of interest. These are sets of imputations, each of which can be obtained from the others by re-labeling the players. Recall that in a symmetric game, only the number of players in a coalition determine the value of the game to the coalition, not who the players are. Therefore a symmetric solution can be "rationalized" without invoking "standards of behavior" which discriminate among the players. Of interest also are *finite* solutions, i.e., those which contain only a finite number of imputations. Clearly such a solution has greater value for a predictive or a normative theory than has a solution with an infinite number of imputations. For example, the $(-1, 0)$ normalized Three-person zerosum game has one finite symmetric solution, namely

$$K = \{(-1, 1/2, 1/2), (1/2, -1, 1/2), (1/2, 1/2, -1)\}, \tag{15.15}$$

which states that two players will take one unit of payoff from the third and will divide it equally. The solution does not state which player will be the loser, but it ought not be expected to do so, in view of the fact that the players are not distinguishable as individuals. This solution is somehow intuitively more "acceptable" than others. It is also finite, hence more specific than non-finite solutions.

Consider now the Eight-person symmetric game given by the following characteristic function:

$$v(\bar{1}) = -1;\ v(\bar{2}) = -2;\ v(\bar{3}) = 1;\ v(\bar{4}) = 0;$$
$$v(\bar{5}) = -1;\ v(\bar{6}) = 2;\ v(\bar{7}) = 1;\ v(\bar{8}) = 0. \tag{15.16}$$

The pro-rata gains are

$$g_1 = -1;\ g_2 = -1;\ g_3 = 1/3;\ g_4 = 0;\ g_5 = -1/5;\ g_6 = 1/3;\ g_7 = 1/7;\ g_8 = 0. \quad (15.17)$$

The largest pro-rata gain is 1/3, and the smallest coalition which can get it contains three players. Therefore $a = 1/3$, $p = 3$. Since $n = 8$, $m = 2$, $r = 2$.

We can now verify that the vector set k consisting of

$$k \equiv \{1/3, 1/3, 1/3, 1/3, 1/3, 1/3, -1, -1\} \quad (15.18)$$

together with 27 others obtained by permuting the payoffs constitute a (finite symmetric) solution. This solution says that six of the players will divide the value accruing to a six-person coalition (i.e., 2) equally among them, the remaining two being the losers.

Now, one might ask why should not a solution consist of imputations in which the *minimal* number of players who can get the maximal pro-rata gain are the winners, while the remaining players are the losers? That is, imputations of the form

$$K' = \{(1/3, 1/3, 1/3, -1/5, -1/5, -1/5, -1/5, -1/5)\} \quad (15.19)$$

and its permutations. We see that this set does not constitute a solution in the sense of Von Neumann and Morgenstern, because the imputations in it dominate each other. For example,

$$\vec{x} = (1/3,\ 1/3,\ 1/3,\ -1/5,\ -1/5,\ -1/5,\ -1/5,\ -1/5) \quad (15.20)$$

dominates

$$\vec{y} = (-1/5,\ -1/5,\ -1/5,\ -1/5,\ -1/5,\ 1/3,\ 1/3,\ 1/3) \quad (15.21)$$

via coalition $(\overline{123})$. No such mutual domination can occur among the imputations of K, because, since $v(\overline{2}) = -2$, no two players can get more than they can get in K.

Note, further, that of any three players, at least one gets 1/3 in any imputation of K, and hence must be indifferent between these imputations.

One might think that this result implies that the Von Neumann-Morgenstern theory is at loggerheads with some ideas expressed by observers of coalition formation, to the effect that *minimal* winning coalitions tend to form. If by a "winning coalition" we mean a coalition in a *simple* game which commands value 1 (cf. Chapter 14), then there is no contradiction, since in that case a minimal winning coalition awards the largest pro-rata gain. If, however, by "minimal winning coalition" one means the coalition with the smallest number of members which awards the largest pro-rata gain in a symmetric game, which is *not* simple, then the Von Neumann-Morgenstern solution is at variance with the conjecture. For, as we have seen, the "minimal winning coalitions" in this latter sense are coalitions with three members in the game just described, while the number of players among whom the positive payoff is split is six.

It should be kept in mind that the imputations in K are by no means undominated. They are not dominated by other imputations in K, but are certainly dominated by some imputations not in K. For example, the two losers and a winner can always go off and form a Three-person coalition by themselves, in which the winner gets 1/2 and the two losers 1/4 each, to the advantage of each of these three. But then there will be another imputation in K which will dominate the new one. As we have already pointed out, the "stability" of the imputations in a Von Neumann-Morgenstern solution is interpreted in terms of the indifference of subsets of players among the imputations in K, and in terms of a tendency to "return" to K after departures from it. Considerations of what happens as a result of such departures and returns underlie W. Vickrey's concept of "strong solution" (see Note 13).

Quota Games

Consider all the two-player coalitions of an N-person game. We shall write v_{ij} for $v(\overline{ij})$, the value assigned by the characteristic function of the game to each pair-coalition (player i with player j) for all $i \neq j$. Suppose we find a vector

$$\vec{\omega} = (\omega_1, \omega_2, \ldots, \omega_n), \quad (15.22)$$

whose components satisfy the following equations:

$$\omega_i + \omega_j = v_{ij} \text{ for all } i \neq j; \quad (15.23)$$

$$\sum_{i=1}^{n} \omega_i = v(N). \quad (15.24)$$

Suppose now $n > 3$, and there is another vector

$$\bar{\omega}' = (\omega'_1, \omega'_2, \ldots, \omega'_n) \quad (15.25)$$

which satisfies equations analogous to (15.23) and (15.24).

Then we must have

$$\omega_1 + \omega_2 = v_{12} = \omega'_1 + \omega'_2. \quad (15.26)$$

$$\omega_2 + \omega_3 = v_{13} = \omega'_2 + \omega'_3. \quad (15.27)$$

Subtracting (15.27) from (15.26), we have

$$\omega_1 - \omega_3 = \omega'_1 - \omega'_3; \quad (15.28)$$

but also

$$\omega_1 + \omega_3 = \omega'_1 + \omega'_3. \quad (15.29)$$

Adding (15.28) and (15.29), we have

$$2\omega_1 = 2\omega'_1; \ \omega_1 = \omega'_1. \quad (15.30)$$

Similarly, all the components of ω can be shown to be equal to the corresponding components of ω'. Therefore the vector ω, if it exists at all, must be unique.

Games for which a vector ω can be found are called

quota games, and the corresponding vector ω is called the quota.

We can quickly find out whether a Three-person game is a quota game. There are three pair-coalitions, v_{12}, v_{13}, and v_{23}. Let

$$\omega_1 = \tfrac{1}{2}(v_{12} + v_{13} - v_{23}) \qquad (15.31)$$

$$\omega_2 = \tfrac{1}{2}(v_{23} + v_{12} - v_{13}) \qquad (15.32)$$

$$\omega_3 = \tfrac{1}{2}(v_{13} + v_{23} - v_{12}). \qquad (15.33)$$

Then $\omega_1 + \omega_2 = v_{12}$; $\omega_1 + \omega_3 = v_{13}$; $\omega_2 + \omega_3 = v_{23}$, so that condition (15.23) is satisfied. Condition (15.24) is satisfied if and only if

$$\tfrac{1}{2}(v_{12} + v_{13} + v_{23}) = v(\overline{123}), \qquad (15.34)$$

which becomes the determining condition for a Three-person game to be a quota game.

We see at once that a constant-sum Three-person game cannot be a quota game. This is because in a (normalized) Three-person game $v_{ij} = v(\overline{123})$ for all $i \neq j$; hence, $v_{12} + v_{13} + v_{23} = 3v(\overline{123})$ instead of $2v(\overline{123})$ as required.

The condition for an arbitrary N-person game to be a quota game turns out to be

$$v_{ij} + v_{k\ell} = v_{ik} + v_{i\ell} \qquad (15.35)$$

where i, j, k, ℓ, are any four distinct players, and

$$\sum_{i \neq j} v_{ij} = 2(n - 1)v(N). \qquad (15.36)$$

In that case,

$$\omega_i = \tfrac{1}{2}(v_{ij} + v_{ik} - v_{jk}) \qquad (15.37)$$

for all distinct players i, j, k.

Quota games with an even number of players have some interesting properties. Whenever sets S, S − T, and $S \cap T$ have an even number of elements,

$$v(S \cap T) + v(S - T) = v(S). \qquad (15.38)$$

Now it follows from the super-additive property of the characteristic function that for all S, T, $\subset$N, the following inequality holds:

$$v(S \cap T) + v(S - T) \geqslant v(S). \qquad (15.39)$$

If n is even and the game is a quota game, inequality (15.39) reduces to an equality. Conversely, if n, $|S|$, $|S \cap T|$, and $|S - T|$ are all even and equality (15.38) holds, the game is a quota game. A consequence of (15.38) is that

$$\sum_{i \in S} \omega_i = v(S) \text{ for all } S \subset N \text{ if } |S| \text{ is even.} \qquad (15.40)$$

In other words, all coalitions with an even number of players get their exact value if each player in a quota game is assigned his quota.

If a Four-person game is constant-sum, the "even" coalitions are the six pairs and the grand coalition. Moreover, if S and T are distinct coalitions, $|S \cap T|$ is even if and only if either $(S \cap T) = \emptyset$ (in which case S and T are complementary); or $S \cap T = S$ (in which case $S = N$); or $S \cap T = T$ (in which case $T = N$). In all of these cases (15.38) is satisfied. Therefore all Four-person constant-sum games are quota games.

Another consequence is that all inessential games are quota games with $\omega_i = v_i$, since, by definition of an inessential game, we must have equation (15.23) satisfied.

Now, although the components of the vector $\vec{\omega}$ (if it exists) add up to v(N), this vector is not necessarily an imputation, because it may turn out that for some player $\omega_i < v_i$, which violates the condition that, in an imputation, every player must get at least his value.

A player who gets less than his value in a quota is called a *weak* player. Clearly, no two players can be weak. For, if players i and j both were weak, we would have

$$\omega_i < v_i; \omega_j < v_j; \omega_i + \omega_j = v_{ij} < v_i + v_j \qquad (15.41)$$

which violates the super-additive property of the characteristic function.

Moreover, if n is odd, there can be no weak player in the corresponding N-person quota game. For, if n is odd,

$$v(N) - v_i \geqslant v(N - \{i\}) \geqslant \sum_{j \neq i} v_j = v(N) - \omega_i. \quad (15.42)$$

Therefore $v_i \leqslant \omega_i$, and player i cannot be weak.

Consider now a quota game, Q, which has no weak player. Then the quota, i.e., the vector $\omega = (\omega_i, \omega_2, \ldots, \omega_n)$, is an imputation, and $c_i \equiv \omega_i - v_i \geqslant 0$ $(i = 1, 2, \ldots, n)$. Define the vector $\vec{\gamma}_{ij}$ as follows:

$$\vec{\gamma}_{ij} = (\omega_1, \omega_2, \ldots v_i, \ldots \omega_j + \omega_i - v_i, \ldots, \omega_n), \quad (15.43)$$

where all the components are the components of ω except the i-th and j-th components. Namely, the i-th component is the value to player i and the j-th component is the quota to player j increased by the excess of player i's quota over his value. In other words, the vector $\vec{\gamma}_{ij}$ is constructed from the quota by taking the "excess" of player i $(\omega_i - v_i)$ away from him and giving it to player j. We shall, therefore, refer to player j as the *beneficiary* of i, and to i as the *benefactor* of j.

There is an interesting connection between the theory of ψ-stability and that of the quota games. Recall that the function $\psi(\mathfrak{T})$ assigned to each partition $\mathfrak{T}$ a set of partitions which involved permissible shifts of players from the coalitions of $\mathfrak{T}$ to other coalitions. Consider now a particular function $\psi^1(\mathfrak{T})$ which permits shifts of no more than one player at a time into or out of any coalition. A game will be called ψ^1-stable if it has at least one ψ^1-stable pair; that is, if there exists some partition $\mathfrak{T}$ and some imputation $\vec{x}$ such that none of the permissible shifts will benefit any of the possible new coalitions. It turns out that a quota game is stable if and only if it has no weak player.

Now, recall that all Four-person constant-sum games are quota games. Such a game may have a weak player; for example, the game given by the characteristic function

$$v(i) = 0 \quad (i = 1, 2, 3, 4) \tag{15.44}$$

$$v_{12} = 0; v_{13} = 1; v_{14} = 1; v_{23} = 3; v_{24} = 3; v_{34} = 4 \tag{15.45}$$

$$v(\overline{ijk}) = 4; v(N) = 4, \tag{15.46}$$

which has a quota $(-1, 1, 2, 2)$. The reader may verify that this game has no ψ^1-stable pair.

On the other hand, if there is no weak player, so that the quota awards to each player at least his value, then clearly the pair $[\omega, \mathfrak{T} = \{(1), (2), (3), (4)\}]$ is stable. The only changes in the coalition structure allowed by $\psi^1(\mathfrak{T})$ are combinations of pairs of players. But there is no inducement for any pair to combine, because in the quota each pair already gets the value to their coalition, since $\omega_i + \omega_j = v(\overline{ij})$ for all $i \neq j$.

There may, however, be other stable pairs, for example, those where the players are paired off and each pair receives its value.

It is tempting to view the quota as a "solution" of games which are quota games. The rationale is as follows: if each player is awarded his quota, then no two players can, by forming a pair coalition, command jointly a larger value than they command separately. In other words, quota games are "pair-wise inessential." One might then argue that, since there is no inducement to form pair coalitions (if each player gets his quota), no coalitions will form at all (assuming that coalitions can form by adding a player at a time), and consequently each player will accept his quota.

This argument is weak for two reasons. First, it ignores

the advantages which may accrue to the coalitions with more than two players. If the players are "rational," they will take these advantages into account. Second, it does not explain the anomaly of the "weak player" whose quota is less than what he is guaranteed if he plays against all others. Hence, in this case, at least the quota cannot be a "solution" (in the sense of an individually rational payoff configuration).[35]

In Chapter 17 we shall have the opportunity to compare the experimentally observed payoff configurations with the quotas in quota games.

16. *Coalitions and Power*

People join in coalitions to increase their power vis-à-vis other people with conflicting interests. Real life examples of coalitions are all too familiar: trade unions, political parties, cartels, military alliances. The cliché "In union there is strength" is expressed in the game-theoretic axiom regarding the super-additivity of the characteristic function:

$$v(S \cup T) \geqslant v(S) + v(T) \quad \text{if} \quad S \cap T = \emptyset. \quad (16.1)$$

In this way, $v(S)$ could be taken as a measure of power of the coalition S. The measure seems natural enough: $v(S)$ is the amount which the coalition S is able to guarantee to its members in the worst circumstance it can face, namely, being confronted by the counter-coalition $-S$, which is determined to keep the joint payoff to S down to its minimum.

This is the worst that can happen to the *coalition,* but it is not the worst that can happen to its *members.* For, as we have seen, there are pressures operating in the N-person game which may disrupt coalitions. Players may be lured away from coalitions which they have temporarily joined, and this may leave their erstwhile partners in a disadvantageous position. Or players outside a coalition may play off the members of the coalition against each other with unfortunate consequences for the latter. Therefore the power of a coalition ought to be conceived as depending not only on what the coalition can get but also on the cohesion of its members; for example, on their ability to resist tempting offers, calculated to break up the coalition.

If a coalition sticks together, it is, for all practical

purposes, a single player. The characteristic function is based on the assumption that the N-person game reduces to a Two-person game, since if a coalition S of s players forms, the remaining n − s players cannot do better than form the counter-coalition −S. However, in real life, this may not happen. Several coalitions may form; that is, the players of an N-person game may be partitioned into a *coalition structure*

$$\mathcal{S} = \{S_1, S_2, \ldots, S_m\},\ m \leqslant n. \qquad (16.2)$$

Then the N-person game reduces to an M-person game, which can be considered in its own right.

Here we shall examine the consequences of an additional assumption, namely, that there is a *standard of fairness,* which apportions the payoffs among the players of any N-person game (for any given n) in accordance with a given rule applied to the characteristic function of that game. One such standard of fairness may be the Shapley value; another the so-called egalitarian standard, to be presently described; still another may be derived from the particular mores governing the social roles of the players.

The Shapley value was discussed in Chapter 5. The egalitarian standard assigns to each player i, in addition to $v(\bar{i})$, a bonus which amounts to an equal share of the total utility gains that accrue to the grand coalition when the individual players merge to form it. Thus, if the value of the game to player i is v(i) and to the grand coalition v(N), the egalitarian standard assigns to player i the amount

$$\varphi_i = v(i) + \frac{v(N) - v(1) - v(2) - \cdots - v(n)}{n}. \qquad (16.3)$$

To illustrate a standard of fairness derived from prevailing social mores, let us suppose that if the sale of a house involves a seller, a buyer, and an agent, then of the total increment of utility which results from the sale, the agent is "entitled" to 10 percent, while the seller and the buyer are entitled to 45 percent each. If, however, two of them are in a coalition, the standard of fairness applies to the way the total utility gain is split between the coalition and the third party.

Suppose, for example, the agent joins either the seller or the buyer in a coalition. In the eyes of society, the combination may be viewed as a "reputable real estate firm," which, as such, is entitled to 60 percent of the utility gain that accrues from the sale of a house. Or suppose, on the contrary, that the seller and the buyer have joined in a coalition. In this case, it may be considered fair to give the agent 5 percent for his trouble and no more, since the two have no further need of him.

In the situation just depicted it was to the advantage of each coalition to enter the game as a single player. The standard of fairness may be such, however, that it is to the advantage of the coalition to enter the game as two or more players. As an example, consider a game given by the characteristic function

$$v(\overline{12}) = 100 = v(\overline{13}); v(\overline{123}) = 120; v(\overline{i}) = 0 \quad (i = 1, 2, 3)$$

$$v(\overline{23}) = 0. \qquad (16.4)$$

Assume that the egalitarian standard of fairness is in force. Although the value of the game to coalition ($\overline{23}$) is zero, the standard of fairness saves the two players, if they enter the game as a single player. For this standard awards to this "player" a payoff of what has become a Two-person game, namely

$$\varphi_{23} = v(\overline{23}) + \frac{v(\overline{123}) - v(\overline{23}) - v(\overline{1})}{2} = 60, \qquad (16.5)$$

and to player 1

$$\varphi_1 = v(1) + \frac{v(\overline{123}) - v(\overline{1}) - v(\overline{23})}{2} = 60. \qquad (16.6)$$

Since the positions of players 2 and 3 are identical in this game, they split the 60 equally between them, and consequently, the disbursement becomes (60, 30, 30).

Actually, players 2 and 3 can do even better if, after having formed their coalition, they enter the game as *two* players. Then the egalitarian standard of fairness will award the following payoffs to the respective players of what is once again a Three-person game:

$$\varphi(\bar{i}) = \frac{v(\overline{123}) - v(\bar{1}) - v(\bar{2}) - v(\bar{3})}{3} = 40 \quad (i = 1, 2, 3). \tag{16.7}$$

We see that under these conditions a coalition which has no power (if judged by the characteristic function) can get up to two-thirds of the total return to the grand coalition, if they enter the game as *two* players.

This model is generalized as follows. The givens are the characteristic function of the game and a standard of fairness. The latter assigns payoffs to certain subsets of players called *pressure groups.* The assignment is in accordance with a specified rule. If the players partition themselves into pressure groups in accordance with the partition

$$\mathcal{P} = \{P_1, P_2, \ldots, P_m\}, \tag{16.8}$$

then the disbursement of payoffs is calculated on the basis of the characteristic function and a function $\varphi(\mathcal{P})$ derived from the standard of fairness. As a result we have a disbursement vector

$$\vec{\varphi}(\mathcal{P}) = \{\varphi_1(\mathcal{P}), \varphi_2(\mathcal{P}), \ldots, \varphi_m(\mathcal{P})\}, \tag{16.9}$$

whose components are functions of the partition $\mathcal{P}$ (a set of sets). The characteristic function enters via restrictions on the function $\varphi(\mathcal{P})$, namely:

$$\varphi_i(\mathcal{P}) \geqslant v(P_i) \tag{16.10}$$

$$\varphi_1(\mathcal{P}) + \varphi_2(\mathcal{P}) + \cdots \varphi_m(\mathcal{P}) = v(N). \tag{16.11}$$

These restrictions will be recognized as the principle of individual rationality and the principle of collective rationality (cf. Chapter 3), applied to the *pressure groups*, acting as individual players.

Suppose now a coalition S forms. The question to be decided by the members of the coalition is whether to enter the game as a single player or as several "players" (i.e., pressure groups). In the meantime, the members of the counter-coalition (if it forms) are trying to decide the same question. Now a coalition S with s members can partition itself in $\gamma(s)$ different ways. The counter-coalition with $n - s$ members can partition itself in $\gamma(n - s)$ different ways where γ is a certain function discussed in combinatorial analysis. Depending on the particular pair of partitions chosen, the function $\varphi(\mathcal{P})$ will assign a payoff to each pressure group, hence a joint payoff to each of the two coalitions, S and $-S$. Call each partition a strategy (since it involves a decision by the corresponding coalition). Therefore the coalitions S and $-S$ are essentially playing a Two-person game in normal form (cf. p. 58). Moreover, the game is constant-sum, since for all $\mathcal{P}$ (i.e., for all choices of strategies)

$$\Sigma\varphi_i(\mathcal{P}) = v(N). \tag{16.12}$$

The solution of this game is given by Two-person game theory. It assigns an optimal strategy (pure or mixed) to each coalition. Corresponding to these strategies a disbursement of payoffs (or expected payoffs) is determined.

Example

Let a Four-person game G be given by its characteristic function v( ). In particular, let

$$v(\bar{i}) = 0;\ v(\overline{12}) = v(\overline{34}) = 1;\ v(\overline{1234}) = 8. \tag{16.13}$$

The remaining values of v( ) are not of interest to us for the moment. We shall assume that coalition $(\overline{12})$ has formed and that the remaining players have formed the counter-coalition $(\overline{34})$. Now each coalition has a choice of entering the game either as one player or as two players. We assume further that the following (non-egalitarian) standard of fairness is in force. Coalition $(\overline{12})$ is entitled to 1/3 of the joint utility gain if $(\overline{12})$ and $(\overline{34})$ merge. If players 1 and 2 enter as separate players, each is entitled to 2/7 of the joint utility gain, while the coalition $(\overline{34})$ takes 3/7. If, on the other hand, players 3 and 4 enter separately, $(\overline{12})$ is entitled to 4/7 of the joint utility gain resulting from the merger. Finally, if all players enter separately, 1 and 2 are entitled to 1/8 each, while 3 and 4 are entitled to 3/8 each. This example may seem artificial, but it may well reflect some situation where the differences in the benefits accruing from the mergers derive from differential costs of effecting the merger, the differential status "in the eyes of society" of the corresponding pressure groups, or what not.

We label the four possible partitions of the players as follows:

$$\mathcal{P}_1 = \{(\overline{12}), (\overline{34})\} \tag{16.14}$$

$$\mathcal{P}_2 = \{(\overline{1}), (\overline{2}), (\overline{34})\} \tag{16.15}$$

$$\mathcal{P}_3 = \{(\overline{12}), (\overline{3}), (\overline{4})\} \tag{16.16}$$

$$\mathcal{P}_4 = \{(\overline{1}), (\overline{2}), (\overline{3}), (\overline{4})\}. \tag{16.17}$$

According to the rule stated above, if $\mathcal{P}_1$ obtains, players 1 and 2 get jointly

$$\varphi_{12}(\mathcal{P}_1) = \text{v}(\overline{12}) + 1/3[\text{v}(N) - \text{v}(\overline{12}) - \text{v}(\overline{34})]$$

$$= 1 + \frac{8 - 1 - 1}{3} = 3. \tag{16.18}$$

If $\mathcal{P}_2$ obtains, players 1 and 2 get, in addition to $\text{v}(\overline{1})$ and $\text{v}(\overline{2})$, which are both zero, 4/7 of the joint utility gain:

$$\varphi_1 + \varphi_2 = v(\bar{1}) + 2/7[v(N) - v(\overline{34}) - v(\bar{1}) - v(\bar{2})]$$
$$+ v(\bar{2}) + 2/7[v(N) - v(\overline{34}) - v(\bar{1}) - v(\bar{2})]$$
$$= 4/7(8 - 1) = 4. \quad (16.19)$$

If $\mathcal{P}_3$ obtains, they jointly get

$$\varphi_{12} = v(\overline{12}) + 4/7(8 - 1) = 5. \quad (16.20)$$

Finally, if $\mathcal{P}_4$ obtains, players 1 and 2 jointly get 2 units. Players 3 and 4 get, of course, what is left of v(N) in each case.

Call the act of splitting into two pressure groups by each coalition strategy 1; that of not splitting, strategy 2. Then the constant-sum game played by the two coalitions is represented in normal form by the following matrix, in which the entries are the payoffs to coalition $(\overline{12})$.

		(34)	
		Strategy 1	Strategy 2
$(\overline{12})$	Strategy 1	2	5
	Strategy 2	4	3

This game has no saddle point; hence the solution prescribes a mixed strategy to each player. Namely, coalition $(\overline{12})$ should choose strategies 1 and 2 with probabilities 1/4 and 3/4 respectively; coalition $(\overline{34})$ should choose the two strategies with equal probabilities. If they do so, then the expected value of this game to coalition $(\overline{12})$ is 7/2, and consequently 9/2 to coalition $(\overline{34})$.

So far we have analyzed only the situation when the four players formed two pair coalitions. Given the complete characteristic function and standard of fairness, we can carry out the appropriate analysis for the other possible situations, namely when other coalitions form. In particular, when a three-against-one partition occurs, the lone player has, of course, no choice; but the three-person coalition can enter the game in five different ways. Such a coalition will be actually playing a One-person game. It can simply choose the partition which maximizes its payoff.

Quite generally, in any N-person game we can calculate the *value* which accrues to each coalition playing the Two-person constant-sum game described above. This determines a function on the subsets of players to real numbers. This function turns out to be always super-additive. Therefore, it can serve as a characteristic function of another game G′, the so-called *derived* game of G. The set of players in game G′ is the same as in game G. However, its characteristic function v′( ), calculated in the manner described is, in general, different from v( ). The value v′(S) assigned by this function to each coalition S has been proposed as a definition of the *power* of S. This power depends both on the original characteristic function v( ) and the particular standard of fairness to which the players adhere.

Having obtained a new characteristic function v′( ), we have obtained a new game G′. Now the question naturally arises about what happens if we go through the same process once more, obtaining a game G″, the derived game of G′. (There seems to be no reason to stop with just one derivation.) It turns out that

1) G′ is a constant-sum game (regardless of whether G was a constant-sum game or not).

2) If G is a constant-sum game, the derived game G′ is identical to G (has the same characteristic function).

From (1) it follows that G″ = G′; G‴ = G″, etc., so that nothing new is obtained after the first derivation. Therefore we are justified in defining *the* power of a coalition, rather than a power, relative to a given derivation.

Another way of defining the power of a coalition is by identifying it with the *maximin* payoff accruing to the coalition in the Two-person game described above. Thus the maximin payoff accruing to $(\overline{12})$ in the game represented by the matrix on p. 240 is 3. The maximin payoff accruing to coalition (34) is 4. Power so defined is sometimes called the *Thrall power* (after Robert Thrall, who proposed the definition). If the Thrall power is assigned to every subset of players, the resulting function is again super-additive and thus defines another derived game. If the original game is constant-sum, the game derived by assigning maximin values is also constant-sum and has the same characteristic function. If, however, the original game is not constant-sum, the Thrall derived game may not be constant-sum, and we face the problem of making the definition of power unique. Fortunately this can be done. It turns out that the power of each coalition in a derived game is never smaller than in the original game, regardless of the standard of fairness used in each derivation. Therefore, for each S, we have

$$v(S) \leqslant v'(S) \leqslant v''(S) \leqslant \ldots . \qquad (16.21)$$

On the other hand v(S) can never exceed v(N) of the original game, and the latter value remains constant in all derived games. By a well known theorem of analysis, every sequence like (16.21), which never decreases but is bounded from above, must tend to a limit. This limit can be defined as *the* Thrall power of coalition S.

Shapley Value as a Power Index

Among the various standards of fairness, we have mentioned the Shapley value. Each partition of the players into m pressure groups defines an M-person game with a characteristic function determined by the payoffs assigned to the pressure groups by the original characteristic function (regarding the various combinations of the pressure groups as coalitions). This M-person game has a Shapley value, which distributes the joint payoff to the grand coalition v(N) among the pressure groups (now regarded as individual players). In this way the Shapley value plays the role of a standard of fairness. However, the Shapley value (i.e., the joint Shapley value accruing to a subset of players) can itself be regarded as an index of power of that subset acting as a coalition, and, in particular, as an index of power of each individual player. Let us see what the rationale of this definition might be.

Consider a weighted majority game (cf. p. 217) as a legislature in which the members vote by roll call, as, for example, in the United States presidential nominating conventions. In these conventions, the roll call is always in alphabetical order. We shall suppose, however, that the order of the roll call is unknown. For simplicity, suppose that all possible roll call orders are equiprobable. Then each member (or a delegation which votes as a unit) has a certain chance of casting the *decisive* vote. For example, in the Republican presidential nominating convention of 1964, South Carolina cast the decisive vote, giving Goldwater the majority. The Shapley value of each player is calculated as the probability (given equiprobable orders of voting) that the player will be able to cast the decisive vote. A player in this situation is called the *pivot*. It is, of course, assumed that the player is not a priori committed to vote one way or another, i.e., that in casting the decisive vote he *exercises* power.

It may happen that a player is never able to cast a decisive vote. To take an example, suppose that the players of a Five-person weighted majority game have 12, 6, 6, 4, and 3 votes respectively; 16 votes constitute a majority. For the player with 3 votes to cast a decisive vote, there must be some combination of players, excluding him, which commands 13, 14, or 15 votes. Since there is no such combination, the last player cannot ever cast a decisive vote. In spite of his 3 votes, this player has no power at all (as measured by the Shapley value, which is zero). Thus there may be a discrepancy between the weight of one's vote and one's voting power. In the example cited, the player with about 10 percent of the "votes" has no power.

An example with an opposite bias can also be cited. Suppose that three players in a weighted majority game have 50, 49, and 1 vote respectively. The three possible orders of voting are (50, 49, 1), (49, 1, 50), (1, 50, 49), (1, 49, 50), (49, 50, 1), and (50, 1, 49). Here the player with one vote is the pivot in one case out of six; so is the player with 49 votes. The player with 50 votes is the pivot in the remaining four cases. The distribution of power (Shapley value) is, accordingly, (2/3, 1/6, 1/6). Here the player with only 1 percent of the "votes" has 1/6 of the power.

The Electoral College

It is interesting to calculate the distribution of power defined by Shapley value in the United States electoral college. In particular, it is interesting to see whether some of the smaller states are deprived of power altogether. Note that in the established (alphabetized) order of voting, Alabama, Colorado, and other states which must record their votes early, can never be pivots. We are supposing, however, that all orders of voting are equiprobable. This model is somewhat more realistic as a representation of a free-for-all horse trading in the building of coalitions than of the formal voting procedure. Of course, it is in the former rather than in the latter that political power manifests itself. It turns out that the smallest states with 3 votes each in the electoral college do have some power. Irwin Mann and L. S. Shapley[36] calculated the distribution of power in the electoral college as of November 8, 1960, and March 29, 1961. The latter figures are given here.

There were 537 electoral votes. New York with 43 commanded .084 of the total power, greater by a factor of 1.045 than its relative voting strength. The smallest states (Alaska, Nevada) with 3 votes each commanded .0054 of the total power, smaller by a factor of .971 than their voting strength. Thus the discrepancies between the power indices and the voting strengths are very slightly in favor of the larger states.

The Loose Two-Party System

In real politics, the picture is considerably complicated by the fact that the "player" is not precisely defined. We cannot view the individual member of a legislature or a delegation as a player as long as his decisions are determined in a large degree by the commitment of his delegation or his party. If these commitments were firm, there would be no problem, since in that case we could consider the delegation or the party as a "player." In general, however, these commitments are not absolutely firm. In the U.S. legislatures only seldom do the two parties vote as single blocs. R. Duncan Luce and Arnold Rogow have attempted to capture this fluidity in their calculations of power distribution in the federal legislature, assuming a two-party system with imperfect party discipline.

Constitutionally, the United States federal legislature is defined as a bicameral body. In fact, however, the President, in addition to his executive functions, assumes also a legislative role, since his consent is required to enact a law (except when his veto is over-ridden). Formally, therefore, the federal legislature is a tri-cameral body consisting of the House of Representatives (437 members), the Senate (101 members, including the vice president, who votes in cases of ties), and the Presidency (1 member). This combination constitutes a composite game (cf. p. 214). The winning coalitions are those composed of majorities in both houses plus the President and those composed of at least two-thirds majorities in both houses.

There are essentially two parties, and the President belongs to one of them. Party discipline is not strict. In many instances members of one party "defect" to vote with the other. The President may also vote against his own party, as when exercising the veto.

Let the set C_1 consist of the members of the majority party and C_2 of those of the minority party. Thus $|C_1| + |C_2| = 437 + 101 + 1 = 539$. We designate by C_1' those of the majority party who, like the admiral in *H. M. S. Pinafore,* always answer their party's call and never think for themselves at all. In other words, the members of C_1' are the *diehards* of their party, who never defect; similarly the members of C_2' are a corresponding subset of the minority party. The remaining members are potential defectors designated by sets C_1'' and C_2'' respectively. Thus

$$C_1 = C_1' \cup C_1''; C_2 = C_2' \cup C_2''. \qquad (16.22)$$

The President may or may not be a defector.

Of particular interest are those cases where the defections "make a difference." These are the defections which swell the ranks of the other party to make a two-thirds majority in both houses; and those which, although failing to achieve two-thirds majorities, do produce majorities in both houses which had not been there. Moreover, only those defectors are of interest which barely succeed in producing the changes, all the additional votes being superfluous.

There are now various situations to consider in which

the "balance of power" is held by different potential coalitions. There are, first, four possible situations with regard to the existing majorities:

1a. The majority party has two-thirds majority in both houses, and the President belongs to that party.

1b. Same as 1a, but the President belongs to the minority party.

2a. The majority party does not have two-thirds majority in at least one of the houses, and the President belongs to that party.

2b. Same as 2a, but the President belongs to the minority party.

These four conditions are mutually exclusive. Next we have 12 possibilities with regard to defections.

3(a, b). The President is a diehard or a potential defector.

4(a, b). The majority party with the help of defectors from the minority party do or do not constitute a two-thirds majority in both houses.

5a. The minority party with the help of the defectors from the majority party fail to achieve a majority in at least one of the houses.

5b. They can in this manner achieve a simple majority (but not two-thirds) in both houses.

5c. They achieve a two-thirds majority in both houses.

The twelve conditions result in the various combinations of conditions from (3), (4), and (5) $(2 \times 2 \times 3 = 12)$.

If each of the first four situations could exist together with each of the last 12, we would have $4 \times 12 = 48$ possible situations. However, some of the combinations are incompatible. It turns out that only 36 of the 48 combinations are internally compatible. Each of these 36 combinations results in a distribution of power among the members of a different coalition. As an example, consider the case where

1) The President is a potential defector.

2) The majority party together with the defectors from the other party achieve only a simple majority.

3) The minority party together with the defectors from the other achieve only a simple majority.

Then the following coalitions can enact laws:

$$(C_1 \cup P);\ (C_1 \cup C_2' \cup P);\ (C_2 \cup C_1' \cup P). \quad (16.23)$$

If a legislator is not a member of *all three* of these coalitions, he has no power in this case, because the coalition of which he is not a member does not "need him." Consequently, in this case, the power is distributed among the members of the set which is the *intersection* of the three sets (16.23), namely

$$(C_1 \cup P) \cap (C_1 \cup C_2' \cup P) \cap (C_2 \cup C_1' \cup P) = C_1' \cup P. \quad (16.24)$$

Now we apply the ψ-stability theory (cf. p. 139) which in this case determines the possible changes in coalition structure (via defining the defectors). It turns out that the coalition (16.24), coupled with some distribution of power (which here plays the part of payoffs) among all its members forms a stable configuration. Note that this distribution of power does not mean that this coalition can enact law, but only that its cooperation is necessary to enact a law.

Using this method, we can determine the "power set" for each of the 36 situations. These are shown in Table 1. Note that only one of the 36 situations results in an impasse; i.e., no stable pair, in the sense of ψ-stability, exists. This happens, of course, when none of the possible alliances can achieve a two-thirds majority in both houses, there are not enough defectors from the majority party to form a majority with the minority party, and the President is a diehard of the minority party. Consequently, the President can effectively veto all bills, but his party cannot pass any bill.

TABLE 1 The power distributions for a stable two-party system under the given condition, from Luce and Rogow.[37]

No.	Presidential Defection	Party of President	Size of Party 1 Majority	Size of Party 1 plus Party 2 Defectors	Size of Party 2 plus Party 1 Defectors
1.a.i	Possible	Either	simple	simple	less than majority
ii					simple
iii					two-thirds
1.b.i				two-thirds	less than majority
ii					simple
iii					two-thirds
2.a			two-thirds	two-thirds	less than majority
b					simple
c					two-thirds
3.a.i	Not Possible	1	simple	simple	less than majority
ii					simple
iii					two-thirds
3.b.i				two-thirds	less than majority
ii					simple
iii					two-thirds
4.a.i		2	simple	simple	less than majority
ii					simple
iii					two-thirds
4.b.i				two-thirds	less than majority
ii					simple
iii					two-thirds
5.a		1	two-thirds	two-thirds	less than majority
b					simple
c					two-thirds
6.a		2	two-thirds	two-thirds	less than majority
b					simple
c					two-thirds

Evidence in support or in refutation of the model just described would not be easy to collect. The classification of the diehards and the defectors is, to be sure, only a matter of voting records. However, the *consequences* of belonging or not belonging to the power set, that is, essentially the payoffs of the game, are difficult to determine unambiguously. In what follows, we shall examine another model, also derived from the Shapley value as an index of power.[38] Here the index can be related to hypothesized political behavior.

The French National Assembly

The French National Assembly, in contrast to the American Congress, is partitioned into many political parties. Party discipline varies from party to party, but on the whole it tends to be rather tight in the parties of the Right and Left and considerably looser in the parties of the Center. For example, according to P. Campbell's study cited by W. H. Riker[39] of the 1953 Assembly, the Communist party and the Republicaines Progressistes manifested almost perfect voting discipline during the Pinay regime (March–December, 1952). The Socialists and the Independants d'Outre Mer almost matched this record, and so did the two rightist peasant groups, Action Paysanne and the Paysans d'Union Sociale. In the parties of the Center (e.g., M.R.P. and the Radicals), on the other hand, defections occurred quite frequently. The events of interest here are not the occasional defections in voting but formal changes in party affiliation, which W. H. Riker calls *migrations*. (Such events are rather rare in the United States, but frequent in France.)

It goes without saying that ideological considerations may play an important part in these migrations. On these matters game theory has nothing to say. However, quite aside from such considerations, it is interesting to see how much can be accounted for by a simple-minded hypothesis suggested by N-person game theory, namely that migrations have something to do with the redistribution of power; in particular, they are motivated by attempts to gain power. Are these migrations generally along the "power gradients"?

In what follows, we shall assume that members of political parties are subject to party discipline, specifically, that parties vote as blocs, and, thus the National Assembly can be represented by a weighted majority game G. The Shapley value, interpreted as a power index, assigns a certain proportion P_A of the total power to Party A. Assuming that all the members of the party share equally in the apportionment of power to their party, we designate the power accruing to member i of Party A by

$$P_i = P_A/a, \qquad (16.25)$$

where $a = |A|$. Initially, the distribution of votes among the parties determines a certain majority game G_1. After member i has left Party A to join Party B, the redistribution of votes changed the game G_1 into another game G_2, in which his power changed from P_i to P_i'. Thus there was an increment (positive or negative) in the power of member i, namely $P_i' - P_i$.

In the two years 1953 and 1954, 34 migrations involving 61 members of the National Assembly were observed, and all the corresponding increments were calculated. If our hypothesis is correct, we would expect the algebraic sum of all these increments to be a substantial positive number, or else that a substantial majority of the increments are positive.

It turned out that in only 10 of the 34 cases the increment was positive, and moreover, that the algebraic sum of the increments was negative. If this is evidence, it is evidence against the hypothesis. However, this does not exhaust the possibilities of the model. W. Riker proposed another hypothesis, namely that migrating adds to the pro-rata power of the party migrated into. If so, the migrating members may be rewarded in some way to the limit of this increment. Examining the increments of the parties migrated into, we find them positive in only 11 cases and so find no evidence for this hypothesis.

Finally, Riker proposes still another hypothesis, namely that a member may be motivated to leave his party if he feels that his share of power in it is less than that of an average member of the Assembly. The initial relative power position of member i is given by

$$\frac{P_i - 1/n}{1/n}, \tag{16.26}$$

where $n = 627$, the total number of members of the Assembly. If this quantity is positive, member i is in an advantageous position; if negative, he is in a disadvantageous position. It turned out that of the 61 individuals who migrated in the period under study, 45 were in a disadvantageous position. Moreover, the magnitudes of quantities (16. 26) which were negative were generally much larger than the magnitudes of the positive ones. Here we find some support for the hypothesis that the migrations tend to be of those members who are in a disadvantageous position.

Now it may be that the power index tends to favor the larger blocs (as we have seen in the case of the electoral college). If so, the result just described may be simply a reflection of the fact that it is the members of the smaller parties that tend to migrate. However, if this were the case, and if the migrations were into the larger parties, we would expect the power increments of the migrators to increase in the process. Since this result is not observed, we must conclude either (a) that in the particular weighted majority game representing the National Assembly, the power index does not favor the larger parties or (b) that the migration bias is not into the larger parties. These conjectures can, of course, be checked independently.

The foregoing studies of the political structure and behavior were offered as examples illustrating the uses to which some game-theoretical concepts can be put. It would be a mistake to expect these drastically simplified models to yield results which would impress a political scientist engrossed in the substantive details of his subject. In particular, the total excision of political content from models which purport to describe political behavior must appear bizarre to the traditional political scientist. It suggests a trivialization of political theory, a charge often leveled against the "behaviorist" school of political science. In response to this charge, two points can be made. First, it is not unreasonable to assume that the political process is a mixture of "content" politics and a "pure" power game. If so, focusing attention on the latter does not mean a denial of the importance of the former. On the contrary, if we can somehow "factor out" the pure power game, we would have a better idea of its relative importance. Eventually we could even pose questions of significance to "evolutionary" political science; for example, whether the importance of the power game in the political process is increasing or decreasing or remains an invariant component in political behavior. Second, the heuristic values of these studies should be apparent. They call attention to the way quantitative methods of theory construction can be used in novel ways. One should not expect substantive results from a fledgling theory. A respectable theory relevant to the concerns of behavioral scientists can come into being only as a result of many years of sustained and systematic efforts.

17. *Experiments Suggested by N-Person Game Theory*

In this chapter we shall discuss some results obtained in experiments designed to test the usefulness of the various concepts developed in N-person game theory.

Constant-sum Four-Person Games

The experiments to be presently described follow a format typically designed for N-person games in characteristic function form. The subjects (players) are instructed by being informed of the characteristic function of the game they are to play. That is, they are told that they may form coalitions among themselves, a coalition being an agreement among its members on how to split the payoff which will accrue to them jointly. This payoff is the value of the game to the coalition as given by the characteristic function of the game. The subjects are further instructed to be guided only by a goal of getting individually as much as they can get. Their bargaining leverage derives, of course, from their freedom to join one or another of the coalitions. In the following experiments all possible coalitions were allowed.

The characteristic functions of the four constant-sum Four-person games used in these experiments are given in Table 2.

It will be noted that Games 1 and 4 are strategically equivalent. This is because $v'(\ )$, the characteristic function of Game 4, can be obtained from $v(\ )$ the characteristic function of Game 1, by the transformation

$$v'(S) = 3v(S)/2 + \sum_{i \in S} a_i \qquad (17.1)$$

where $a_1 = -20$, $a_2 = -40$, $a_3 = -40$, $a_4 = -20$.

Similarly, Games 2 and 3 are strategically equivalent. Note that Game 3 is a symmetric game. Therefore, although Game 2 does not *look* symmetric, it is nevertheless also a symmetric game.

Each of these games was played eight times by the subjects, who were constantly rotated as players to prevent fixation of outcomes. The idea was to compare the average payoffs received by each player (whose role was assumed by different subjects in different plays) with some theoretical values. The rotation of the subjects in the roles of players presumably "washed out" fluctuations due to individual differences. In my opinion, this is the appropriate approach (at least as a beginning) to experiments based on the idea of game theory. Since

TABLE 2 [40]

Coalition	v(s)			
	Game 1	Game 2	Game 3	Game 4
$(\bar{1})$	0	−40	−20	−20
$(\bar{2})$	0	10	−20	−40
$(\bar{3})$	0	0	−20	−40
$(\bar{4})$	0	−50	−20	−20
$(\overline{12})$	60	10	0	30
$(\overline{13})$	40	0	0	0
$(\overline{14})$	20	−50	0	−10
$(\overline{23})$	60	50	0	10
$(\overline{24})$	40	0	0	0
$(\overline{34})$	20	−10	0	−30
$(\overline{123})$	80	50	20	20
$(\overline{124})$	80	0	20	40
$(\overline{134})$	80	−10	20	40
$(\overline{234})$	80	40	20	20

game theory has nothing to say about the effects upon outcomes of *personal* characteristics of individual players, it makes sense to design experiments in such a way as to minimize the effects of these characteristics and bring out the effects due to the *roles* of the individuals as players in the various games.

Let us now compare the average payoffs accruing to players 1, 2, 3, and 4 in the four games with the respective Shapley values (Tables 3 and 4).[41]

TABLE 3 Payoffs to each of the four players in eight plays of Game 1. The last column lists the coalitions in the order in which they were observed to form. Thus (34), (234) means that coalition (34) formed first, then player 2 joined it. Consequently, in this instance, player 1 was left by himself.

	Play No.	Players 1	2	3	4	Coalitions in the order of formation
	1	0	32	24	24	$(\overline{34});(\overline{234})$
	2	0	34	34	12	$(\overline{23});(\overline{234})$
	3	10	30	30	10	$(\overline{23});(\overline{14})$
	4	10	35	35	0	$(\overline{23});(\overline{123})$
	5	20	40	10	10	$(\overline{12});(\overline{34})$
	6	34	34	0	12	$(\overline{124})$
	7	15	35	30	0	$(\overline{23});(\overline{123})$
	8	35.5	35.5	0	9	$(\overline{12});(\overline{124})$
Average payoff		15.6	34.4	20.3	9.6	
Shapley value		20	26.7	20	13.3	
Quota		20	40	20	0	
No. of times player was in coalition		6	8	6	6	

TABLE 4 Payoffs to each of the four players in eight plays of Game 4. The last column lists the coalitions in the order in which they were observed to form. Thus $(\overline{13})$, $(\overline{134})$ means that coalition $(\overline{13})$ formed first, then player 1 joined it. Consequently, in this instance, player 2 was left by himself.

	Play No.	Players				Coalitions in the order of formation
		1	2	3	4	
	1	25	−40	−10	25	$(\overline{134})$
	2	−20	10	10	0	$(\overline{23})$;$(\overline{234})$
	3	15	−40	15	10	$(\overline{13})$;$(\overline{134})$
	4	25	25	−40	−10	$(\overline{12})$;$(\overline{124})$
	5	24	24	−40	−10	$(\overline{12})$;$(\overline{124})$
	6	−5	5	5	−5	$(\overline{23})$;$(\overline{14})$
	7	25	25	−40	−10	$(\overline{12})$;$(\overline{124})$
	8	15	−40	10	15	$(\overline{134})$
Average payoff		13.0	−3.9	−11.2	2.1	
Shapley value		10	0	−10	0	
Quota		10	20	−10	−20	
No. of times player was in coalition		7	5	5	8	

Comparing the average payoffs with the Shapley values in Game 1, we see a rough agreement. The player with the largest Shapley value (player 2) also got the largest average payoff. The player with the smallest Shapley value (player 4) got the smallest payoff. Players 1 and 3 with equal Shapley values got the intermediate payoffs which, while not identical, are not too disparate. In Game 4 the picture is similar. The players with the largest and smallest Shapley values got the largest and smallest payoffs respectively. Players 2 and 4 with equal Shapley values got the not too disparate payoffs. Possibly, in an experiment with a large number of plays and players randomly assigned to rotating roles (so as to wash out individual differences), the agreement might be still closer. It looks as though the Shapley value is a fairly good predictor of the payoff disbursements in this instance.

Another interesting question is whether the payoffs of strategically equivalent games transform properly into each other. Recall that all the values of Game 4 are obtained from those of Game 1 by the transformation given by equation (17.1). Consequently the payoffs should also be transformed by the same equation. Applying transformation (17.1) to the average payoffs of Game 1, we obtain the (theoretically predicted) payoff of Game 4, as shown in Table 5.

That is to say, if we interpret the average payoffs observed in Game 1 as the "norm" for that game, we ought to observe the payoffs in Game 4 shown in the

TABLE 5

Payoff	Transformation	Average payoff predicted for Game 4	Average payoff observed in Game 4
$x_1 =$	$\frac{15.6 \times 3}{2} - 20 =$	3.4	13.0
$x_2 =$	$\frac{34.4 \times 3}{2} - 40 =$	11.6	−3.9
$x_3 =$	$\frac{20.3 \times 3}{2} - 40 =$	−9.55	−11.2
$x_4 =$	$\frac{9.6 \times 3}{2} - 20 =$	−5.6	2.1

third column of Table 5. Similarly, if we took the average payoffs observed in Game 4 as the norm, we would expect to observe the average payoffs in Game 1 as shown in Table 6.

TABLE 6

Payoff	Transformation	Average payoff predicted for Game 1	Average payoff observed in Game 1
x_1	$2/3(13.0 + 20)$	22.0	15.6
x_2	$2/3(-3.9 + 40)$	24.1	34.4
x_3	$2/3(-11.2 + 40)$	18.8	20.3
x_4	$2/3(2.1 + 20)$	14.7	9.6

In favor of Game 1 as a norm is the zero base line of the individual values, as given by the characteristic function. Then it seems as if players 2 and 3 get less than the amounts expected, while players 1 and 4 get more. The reason for this may lie in the discrepancies in the values of the game to the four players of Game 4:

$$v(\bar{1}) = v(\bar{4}) = -20, \; v(\bar{2}) = v(\bar{3}) = -40. \quad (17.2)$$

In other words, the players' perception of the discrepancies of the *individual* values in Game 4 influences the outcome as well as the Shapley value.

In favor of Game 4 as a norm is the fact that it is a zerosum game, which may make its strategic structure more transparent. If so, then players 2 and 3 are better off in Game 1 at the expense of players 1 and 4. The reason for this may be in the differences between the average values of the coalitions in which the players participate. The average value is 53 (the largest) for player 2 and 27 (the smallest) for player 4. However, the average value of coalitions in which players 1 and 3 participate is 40 in both. The discrepancy in the payoffs accruing to these two players (if it is a significant discrepancy) remains unexplained.

We turn to Games 2 and 3. Game 3 is symmetric, and Game 2 is strategically equivalent to it. The payoffs of Game 3 are obtained from those of Game 2 by the transformation

$$v'(S) = v(S) + \sum_{i \in S} a_i, \qquad (17.3)$$

where $a_1 = +20$, $a_2 = -30$, $a_3 = -20$, $a_4 = +30$.

The results are shown in Tables 7 and 8.[42]

TABLE 7 Game 2—observed payoffs and coalitions.

	Play No.	Players 1	2	3	4	Coalitions
	1	−4	20	30	−46	$(\overline{14});(\overline{23})$
	2	−2	26	26	−50	$(\overline{23});(\overline{123})$
	3	−25	25	25	−25	$(\overline{14});(\overline{23})$
	4	−20	35	35	−50	$(\overline{123})$
	5	−23	25	25	−27	$(\overline{23});(\overline{14})$
	6	−18	25	25	−32	$(\overline{23});(\overline{14})$
	7	−40	42	41	−43	$(\overline{234})$
	8	−40	34	34	−28	$(\overline{234})$
Average payoff		−21.5	29.0	30.1	−37.6	
Shapely value		−20	30	20	−30	
Quota		−20	30	20	−30	
No. of times player participated in coalition		6	8	8	6	

Again we observe that the Shapley value predicts the outcomes rather well, especially in the version where the symmetry of the game is apparent (Game 3). The characteristic function of Game 2 is obtained from that

TABLE 8 Game 3—observed payoffs and coalitions.

	Play No.	1	2	3	4	Coalitions
	1	0	0	0	0	$(\overline{12});(\overline{134})$
	2	−20	1	9	10	$(\overline{34});(\overline{234})$
	3	7	6	7	−20	$(\overline{123})$
	4	10	9	−20	1	$(\overline{12});(\overline{124})$
	5	0	0	0	0	$(\overline{13});(\overline{24})$
	6	−20	9	2	9	$(\overline{24});(\overline{234})$
	7	0	0	0	0	$(\overline{34});(\overline{12})$
	8	0	0	0	0	$(\overline{1234})$
Average payoff		−2.9	3.1	−0.2	0	
Shapley value		0	0	0	0	
Quota		0	0	0	0	
No. of times player participated in coalition		6	8	7	7	

of Game 3 by the same transformation, where $a_1 = -20$, $a_2 = 30$, $a_3 = 20$, $a_4 = -30$. Hence, taking the symmetric-appearing Game 3 as the norm, we obtain the payoffs predicted for Game 2 by the results of Game 3. These are shown in Table 9.

TABLE 9

Player	Average payoff predicted for Game 2	Average payoff observed in Game 2
1	−22.9	−21.5
2	33.1	29.0
3	19.8	30.1
4	−30.0	−37.6

If the characteristic function (as explicitly given) influences the outcomes (psychologically), players 2 and 3 should get more in Game 2 than the payoffs predicted for them, while players 1 and 4 should get less. It turns out, however, that players 1 and 3 get more while players 2 and 4 get less. However, if we compare the last column of Table 9 with the Shapley value and ignore the small discrepancies in the payoffs of players 1 and 2, then the results are in the right direction, at least for players 3 and 4: the former gets somewhat more than the Shapley value; the latter gets somewhat less.

Comparing the average payoffs to the respective players with their quotas (recall that all Four-person constant-sum games are quota games), we note that, where the quotas differ from the Shapley values (in the non-symmetric Games 1 and 4), both predict about equally well in Game 1, but in Game 4 the quota does worse; in particular, it assigns "unrealistic" quotas to players 2 and 4.

In summary, we see that the Shapley value of a constant-sum game is a fairly good base line predictor of the average outcome. We see also that discrepancies brought about by transforming the characteristic functions of such games (while preserving strategic equivalence) can be partially explained by the induced inequalities of the individual values, as given by the transformed characteristic function.

Finally, it is interesting to note that a Three-person coalition failed to form in only three cases out of sixteen, where non-symmetric Games 1 and 4 were played (cf. Tables 10, 11); but it failed to form in seven cases out of sixteen when the (symmetric) Games 2 and 3 were played. In eight cases, two pairs were formed, and in one case the grand coalition. It seems as if the symmetry of the game tends to induce symmetric partitions of the players.

Tests of ψ-stability

The same experimental results may be used to test the ψ-stability theory (cf. Chapter 8). For this purpose, it is convenient to transform the payoffs so that $v(N) = 1$; $v(\bar{i}) = 0$ $(i = 1, 2, 3, 4)$. Then the strategically equivalent Games 1 and 4 become identical, as do the strategically equivalent Games 2 and 3.

The question arises about what restrictions should be placed on coalition changes, as a model with which to compare the results of the experiment. One possibility is to allow any coalition to recruit a single member or else to expel a single member. Under these restrictions, the only stable pairs in the symmetric games are those involving the "every man for himself" partition $\{(1), (2), (3), (4)\}$ and the quota of the game (cf. p. 232). In the non-symmetric games, these are also stable pairs, and, in addition, the partition $\{(123), (4)\}$ can form stable pairs with certain imputations. We see that these predictions are clearly not realized, and therefore the model represented by these restrictions must be discarded.

If the restrictions on changes of coalition structure are more stringent, more stable pairs appear. It turns out that if we strengthen these restrictions only slightly, the experimental results can be accounted for fairly well. Namely, let each coalition be allowed to expel any member as before, but let it be permitted to recruit a member only if he is not already in a coalition with someone else. The comparison of predicted and observed results are shown in Tables 10 and 11.[43] In the first column the stable pairs are given; namely, a partition together with the restrictions on the imputations which form stable pairs with that partition. In the second and third columns the game and the play are indicated, where the partition was observed. In the

TABLE 10

Coalition structure and corresponding restrictions on imputations	Game No.	Play	Observed imputation 1	2	3	4	Observation compatible with prediction?
$\{(1), (\overline{234})\}$							
$x_2 + x_3 \geqslant 0.75$							
$x_2 + x_4 \geqslant 0.50$	1	1	.00	.40	.30	.30	No
$x_3 + x_4 \geqslant 0.25$	1	2	.00	.43	.43	.15	Yes
$x_4 = 0.00$	4	2	.00	.42	.42	.17	Yes
$\{(2), (\overline{134})\}$							
$x_1 + x_3 \geqslant 0.50$	4	1	.38	.00	.25	.38	Yes
$x_1 + x_4 \geqslant 0.25$							
$x_3 + x_4 \geqslant 0.25$	4	3	.29	.00	.46	.25	Yes
$x_2 = 0.00$	4	8	.29	.00	.42	.29	Yes
$\{(3), (\overline{124})\}$	1	6	.43	.43	.00	.15	Yes
$x_1 + x_2 \geqslant 0.75$	1	8	.44	.44	.00	.11	Yes
$x_1 + x_4 \geqslant 0.25$	4	4	.38	.54	.00	.08	Yes
$x_3 + x_4 \geqslant 0.50$	4	5	.37	.53	.00	.10	Yes
$x_3 = 0.00$	4	7	.38	.54	.00	.08	Yes
$\{(\overline{123}), (4)\}$							
$x_1 = 0.25, x_2 = 0.50$	1	4	.13	.44	.44	.00	No
$x_3 = 0.25, x_4 = 0.00$	1	7	.19	.44	.38	.00	No
$\{(\overline{14}), (\overline{23})\}$							
$x_1 + x_4 = 0.25$	1	3	.13	.38	.38	.13	Yes
$x_2 + x_3 = 0.75$	4	6	.13	.38	.38	.13	Yes
$\{(\overline{12}), (\overline{34})\}$							
$x_1 + x_2 = 0.75$							
$x_3 + x_4 = 0.25$	1	5	.25	.50	.13	.13	Yes

fourth column the corresponding observed incompatibilities, if any, between the predicted and the observed outcomes are noted. It must be remembered that, in the context of this theory (unlike the Shapley value), not a specific imputation but only a range of imputations, coupled with a stable coalition structure, is predicted.

We note that the ψ-stability model (assuming the particular function) predicts the outcomes of the symmetric games very well but fails in three cases out of sixteen in the non-symmetric games. The deviation associated with the partition $\{(\overline{1}), (\overline{234})\}$ is not large: $x_2 + x_3 = 0.70 < 0.75$. The deviations associated with

TABLE 11

Coalition structure and corresponding restrictions on imputations	Game No.	Play	Observed imputation 1	2	3	4	Observation compatible with prediction?
$\{(1), (\overline{234})\}$							
$x_2 + x_3 \geqslant 0.50$	2	7	.00	.40	.51	.09	Yes
$x_2 + x_4 \geqslant 0.50$	2	8	.00	.30	.43	.28	Yes
$x_3 + x_4 \geqslant 0.50$	3	2	.00	.26	.36	.38	Yes
$x_1 = 0.00$	3	6	.00	.36	.28	.36	Yes
$\{(3), (\overline{124})\}$							
$x_1 + x_2 \geqslant 0.50$							
$x_1 + x_4 \geqslant 0.50$							
$x_2 + x_4 \geqslant 0.50$	3	4	.38	.36	.00	.26	Yes
$x_3 = 0.00$							
$\{(\overline{123}), (\bar{4})\}$							
$x_1 + x_2 \geqslant 0.50$	2	2	.48	.20	.33	.00	Yes
$x_1 + x_3 \geqslant 0.50$	2	4	.25	.31	.44	.00	Yes
$x_2 + x_3 \geqslant 0.50$	3	3	.34	.33	.34	.00	Yes
$x_4 = 0.00$							
$(\overline{14}), (\overline{23})$							
$x_1 + x_4 = 0.50$	2	1	.45	.13	.38	.05	Yes
	2	3	.19	.19	.31	.31	Yes
$x_2 + x_3 = 0.50$	2	5	.21	.19	.31	.29	Yes
	2	6	.28	.19	.31	.23	Yes
$(\overline{12}), (\overline{34})$							
$x_1 + x_2 = 0.50$	3	1	.25	.25	.25	.25	Yes
$x_3 + x_4 = 0.50$	3	7	.25	.25	.25	.25	Yes
$(\overline{13}), (\overline{24})$							
$x_1 + x_3 = 0.50$	3	5	.25	.25	.25	.25	Yes
$x_2 + x_4 = 0.50$							
$(\overline{1234})$ Does not form a stable pair into any imputation	3	8	.25	.25	.25	.25	*

* Note: the coalition structure $\{(\bar{1}), (\bar{2}), (\bar{3}), (\bar{4})\}$ forms a stable pair with the quota (.25, .25, .25, .25). But the only way the players so partitioned could achieve the imputation is by "declaring" themselves to be the grand coalition. In accordance with the stability of the quota, they would then divide the imputation equally. If the outcome is so interpreted, it is compatible with the model.

the partition $\{(\overline{123}), (\bar{4})\}$ are larger. There player 3 gets more than is allotted to him at the expense of players 1 and 2. An explanation of the discrepancy has not occurred to the author.

From one point of view the predictions derived from ψ-stability models are weaker than those derived from the Shapley value. The latter predicts a unique imputation, while the former, in general, predict only imputations within certain ranges. On the other hand, the ψ-stability models make differential predictions of outcomes, depending on which coalition structure obtains; so that from this point of view its predictions are stronger. This is because, in practice, one would test the Shapley value prediction only against the average outcome of many plays (as was done in this chapter), while the ψ-stability model predictions must be tested separately for each of the observed coalition structures.

Testing the Bargaining Set Model

We shall now examine a set of experiments performed by Michael Maschler on Israeli high school children. The games were Three-person and Four-person non-constant-sum games in generalized (not necessarily super-additive) characteristic function form. The results are shown in Tables 12–15.[44] The first column gives the

TABLE 12

Game No.	12	13	23	123	Outcome	Deviation	Time of negotiation
50	50	100	100	150	50 50 50;G	0 0 0	7 5 2
10	40	80	80	110	30 30 50;G	0 0 0	35
14	60	70	90	150	50 50 50;G	0 0 0	30
19	60	80	100	150	46 50 54;G	0 0 0	10
24	20	30	60	100	20 30 50;G	0 0 0	10
29	30	30	80	100	10 45 45;G	0 0 0	15 5 10
35	0	50	50	100	25 25 50;G	0 0 0	5
36	0	50	50	120	36 36 48;G	0 0 0	20
40	0	50	70	80	5 25 50;G	0 0 0	10 5 5
41	0	50	70	100	20 39 41;G	0 0 0	16

game number as labeled in the experiment; the next four columns give the values to the corresponding coalitions (the values to the individual players being all zero). The outcome, in the sixth column, is designated by the payoff disbursement to the three play-

TABLE 13

Game No.	12	13	23	123	Outcome			Deviation			Negotiation
39	0	50	70	70	2	18	50;G	2	−2	0	10
34	0	50	50	60	15	15	30;G	5	5	−10	10
13	60	70	90	110	30	40	40;G	10	0	−10	3
17	60	80	100	110	10	50	50;G	−20/3	40/3	−20/3	10
18	60	80	100	120	35	35	50;G	15	−5	−10	7
3	50	100	100	110	30	30	50;G	10	10	−20	15 5
4	50	100	100	125	35	35	55;G	10	10	−20	30 10
9	40	80	80	100	32	32	36;G	12	12	−24	30

ers and by the coalition which formed (G denotes the grand coalition). The next column gives deviations, if any, of the actual payoffs from the payoffs allotted in the bargaining set. The last column gives the time in minutes spent in negotiations, where several entries indicate that more than one negotiation session took place.

TABLE 14

Game No.	12	13	23	123	Outcome				Deviation		Time of negotiation
6	40	80	80	0	20	20	0	12	0	0	14 7
8	40	80	80	90	0	20	60	23	0	0	25
20	20	30	60	0	0	28	32	23	0	0	2
21	20	30	60	40	0	25	35	23	0	0	5
22	20	30	60	55	0	23	37	23	0	0	4
25	30	30	80	0	0	40	40	23	0	0	1
26	30	30	80	20	0	40	40	23	0	0	2
27	30	30	80	70	0	40	40	23	0	0	5
28	30	30	80	80	0	40	40	23	0	0	5
30	70	90	120	0	20	0	70	13	0	0	5
37	0	50	70	0	0	12	58	23	0	0	3
38	0	50	70	60	0	15	55	23	0	0	30

The players have been labeled so that 1 is always the "weakest" and 3 the "strongest," unless two of them are of equal strength. (By the "strength" of the player we mean the payoff assigned to him in the bargaining set.)

We have listed the games and their outcomes so that they are grouped as follows. In Table 12 we have games which resulted in the grand coalition and where the payoffs to the players were within the bargaining set (i.e., the deviations from the nearest payoff in the bargaining set is zero for each player). In Table 13 are the games which resulted in the grand coalition where the payoffs deviate from the bargaining set. The games

TABLE 15

Game No.	12	13	23	123	Outcome			Deviation			Negotiation	
16	60	80	100	100	0	43	57	23	3	−3	?	
11	60	70	90	0	0	35	55	23	−5	5	30	5
12	60	70	90	90	0	45	45	23	5	−5	15	10
15	60	80	100	0	25	0	55	13	5	−5	5	
23	20	30	60	60	5	15	0	12	5	−5	20	
7	40	80	80	80	30	0	50	13	10	−10	10	
32	20	30	60	60	10	0	20	13	10	−10	15	
2	50	100	100	100	0	40	60	23	15	−15	40	
31	70	90	120	0	35	35	0	12	10	−15	20	15
33	0	50	50	50	20	0	30	13	20	−20	15	

are listed in increasing order of the absolute magnitude of the largest deviation. In Table 14 are the games which resulted in a coalition of two players and no deviations from the bargaining set. In Table 15 are the games which resulted in coalitions of two players and some deviations from the bargaining set, again in the increasing order of the largest deviation.

Next we note that all of the games in Table 12 are super-additive; that is, in these games the grand coalition gets more than any pair of players. In Table 13 all the games except one (Game 39) are super-additive.

The first gross conclusion to be drawn from these data is that the grand coalition forms *only* if the game is super-additive; in other words, if there is something to be gained collectively by forming the grand coalition.

But does the grand coalition always form if the game is super-additive? To answer this question, let us see whether we can find some super-additive games among those which did not result in the grand coalition, i.e., the games in Tables 14 and 15. We find only one strictly super-additive game, namely Game 8. We also find several non-strictly super-additive games, e.g., Games 28, 7, 12, etc.

We can therefore summarize these results as follows. A grand coalition of three players is likely to form if and only if the game is *strictly* super-additive. On this basis, the principle of collective rationality is satisfied, for there is no reason to include the third player in a coalition if, by joining it, he contributes nothing or "less than nothing" to the coalition. (After all, to make the coalition meaningful, he must be given something, and this must be taken out of the joint payoff of the other two.) Only one exception occurred in Game 39, where player 1 was given 2 units which players 2 and 3 were not obligated to give him, since together they could have gotten all 70 units. These 2 units came out of the payoff of the weaker of the two players (player 2). This outcome may have been due simply to the generosity of player 2, who may have been disturbed by the prospect of excluding the weakest player altogether.

The reader may wish to do an exercise to test his understanding of the kernel theory, by computing the kernels of Maschler's Three-person games and comparing the experimental results with the hypothesis that the observed payoff disbursement shall be in the kernel.

It is interesting to compare the average negotiation times in the four groups of games. These are shown in Table 16.

The interesting difference is between Groups II and IV. In cases where paired coalitions formed, it took longer to arrive at a disbursement which was outside the bargaining

TABLE 16

Groups of Games	I Grand Coalition No deviations	II Pair Coalitions No deviations	III Grand Coal. Deviations	IV Pair Coal. Deviations
Mean negotiation (time in minutes)	18.0	9.0	16.25	23.9

set than a disbursement consistent with the bargaining set.

When we examine the deviations from the bargaining set, we find that, in the case of the weakest player, these are positive in seven cases out of eight when the grand coalition formed; and in the case of the weakest of the two players in the pair coalition, the deviations are positive in nine cases out of ten, where deviations occurred. The average deviation in the payoff of the weakest player in the grand coalition was +5.5; of the weaker player in the pair coalitions, +7.8.

Taking the mean deviations over all games, they are as follows: the average deviation in the payoff of the strongest player is −4.9; in that of the middle player, +3.8; in that of the weakest player, +1.1.

Finally, we are in a position to test the hypothesis, sometimes proposed by behavioral scientists, to the effect that weaker players are more likely to join in a coalition against the strongest player. We note that in fourteen of the twenty-two games, in which a two-person coalition formed, it is the two *stronger* players who joined against the weakest one. In five of the games the weakest joined with the strongest. Only in three of the games did the weaker players (1 and 2) join against the strongest. In this context, the hypothesis is definitely refuted. We shall have more to say on this matter in Chapter 19.

18. *"So Long Sucker": A Do-it-yourself Experiment*

The reader with more than a passing interest in N-person game theory will do well to acquaint himself not only with its formal structure but also with the actual situations which the theory attempts to capture. He can do so by participating in the following game invented in 1950 by M. Hausner, J. Nash, L. Shapley, and M. Shubik.[45] The rules of the game are given below.

1. A Four-person game.

2. Each player starts with 7 *chips* (playing cards or other markers may be used instead), distinguishable by their color from the chips of any other player. As the game proceeds, players will gain possession of chips of other colors. The players must keep their holdings in view at all times.

3. The player to make the first move is decided by chance.

4. A *move* is made by playing a chip of any color out into the playing area, or on top of any chip or pile of chips already in the playing area.

5. The *order of play*, except when a capture has just been made or a player has just been defeated (Rules 6 and 9), is decided by the last player to have moved. He may give the move to any player (including himself) whose color is not represented in the pile just played on. But if all players are represented in that pile, then he must give the move to the player whose most-recently-played chip is furthest down in the pile.

6. A *capture* is accomplished by playing two chips of the same color consecutively on one pile.[46] The player designated by that color must kill one chip, of his choice,

out of the pile, and then take in the rest. He then gets the next move.

7. A *kill* of a chip is effected by placing it in the "dead box."

8. A *prisoner* is a chip of a color other than that of the player who holds it. A player may at any time during the game kill any prisoner in his possession, or *transfer* it to another player. Such transfers are unconditional and cannot be retracted. A player may not transfer chips of his own color, nor kill them, except out of a captured pile (Rule 6).

9. *Defeat* of a player takes place when he is given the move and is unable to play, through having no chips in his possession. However, his defeat is not final until every player holding prisoners has declared his refusal to come to the rescue by means of a transfer (Rule 8). Upon defeat, a player withdraws from the game, and the move *rebounds* to the player who gave him the move. (If the latter is thereby defeated, the move goes to the player who gave *him* the move, and so on.)

10. The chips of a defeated player remain in play as prisoners, but are ignored in determining the order of the play (Rule 5). If a pile is captured by the chips of a defeated player, the entire pile is killed, and the move rebounds as in Rule 9.

11. The *winner* is the player surviving after all others have been defeated. Note that a player can win even if he holds no chips and even if all chips of his color have been killed.

12. *Coalitions*, or agreements to cooperate, are permitted, and may take any form. However, the rules provide no penalty for failure to live up to an agreement. Open discussion is not restricted, but players are not allowed to confer away from the table during the game, or make agreements before the start of the game.

The first question that arises is whereof consists the "skill" of playing this game. "Winning" involves the defeat of all other players. Players can be defeated, however, only by coalitions of other players. Clearly, then, the first prerequisite of skill is the ability to avoid being "ganged up on" by other players. The initial situations of the players are exactly the same. How does one avoid being an odd man when the coalitions are formed? If one becomes an odd man, how does one entice another player away from the coalitions, and, having enticed him, how does one keep his "loyalty"? Finally, when is it of advantage to betray the confidence of a coalition partner? Note that the rules of the game explicitly permit such betrayals, whence the game gets its name. Therefore, assuming that the "ethics" of the game are embodied entirely in its rules (as in most parlor games), ethical considerations do not enter *directly* into decisions. They may, however, enter indirectly; that is, one may be "ethical" as a means to an end. As we shall presently see, successful coalition strategies depend on mutual trust among the coalition partners. The structure of the game is such that *eventually* mutual trust must give way to suspicion, because as soon as a player outside a coalition is defeated, the members of the coalition become potential opponents. It turns out that being completely trustworthy practically insures one's eventual defeat. On the other hand, breach of confidence by a player encourages the formation of coalitions against him. What is the strategically optimal balance between ruthlessness and dependability? *Is* there such a balance in any meaningful sense?

We shall briefly consider some strategic aspects of this game. Call the players Spades, Clubs, Diamonds, and Hearts, and suppose that the first move falls to Spades. It will not do for Spades to form a coalition with just one player. For suppose he forms a coalition with Clubs. His first move obliges him to put a spade on the table. If Spades and Clubs give the move back and forth between them, they can make captures, but they will be capturing their own cards and killing some in the process, thus accomplishing nothing but depleting their resources. Suppose, then, that Spades after playing a spade gives the move to Diamonds. The opposing coalition can now capture his spade. Namely, Diamonds plays a diamond on top of the spade and gives the move to Hearts (Rule 5). Hearts now plays a heart *separately*, and this gives him the right to give the move back to Diamonds. Diamonds now plays another diamond on top of the previous one, captures the pile, and kills the spade.

It is now Diamonds' move. Hence he is in the same position that Spades was in at the beginning of the game. However, he still has a full complement of diamonds, while Spades has one spade fewer. Also Hearts has one of his cards in the playing area. If the game continues with the roles of Spades and Diamonds interchanged, and also with the roles of Hearts and Clubs interchanged, we see that, after the next capture, Spades and Diamonds will have lost a card each to the dead file, while Hearts and Clubs will have one card each in the playing area, and the move will again be Spades'.

Let us pursue the game further, assuming for simplicity that each player started with only three cards. The situation just depicted is shown in Figure 17.

Spades now has a choice of putting a spade on the club, on the heart, or separately. Assume that he is still in partnership with Clubs. He will, therefore, place his card on the "hostile" rather than on the "friendly" card. It will not do to pass the move to the Red Coalition, since between them, they can capture the pile. Spades, therefore, gives the move back to Clubs. Clubs takes in his own card, which he must kill. The situation now is as depicted in Figure 18.

The move is Clubs'. If he places a club on the pile, he must give the move to Diamonds. Diamonds can then put a diamond on the same pile and give the move to Hearts. Hearts can play a heart aside, return the move

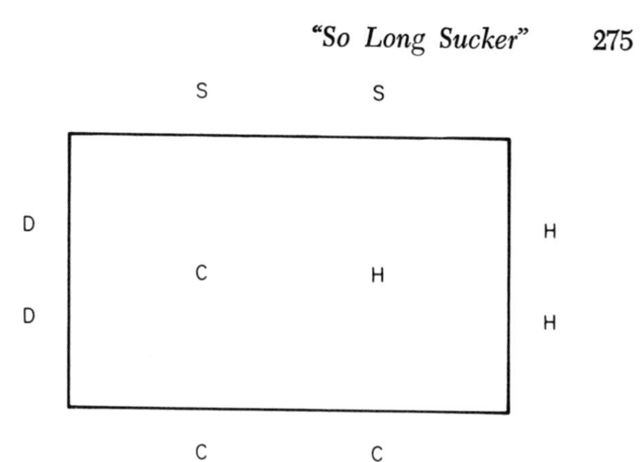

Fig. 17. The box is the playing area. The letters surrounding it designate cards in the possession of players.

to Diamonds, who will then capture the pile. The Red Coalition will be ahead. If Clubs is still faithful to the Black Coalition, he will not play his club on the pile. He must, therefore, put it aside. If he now calls on Diamonds or Hearts, either of them can capture the club by passing the move back and forth. So he calls on his partner

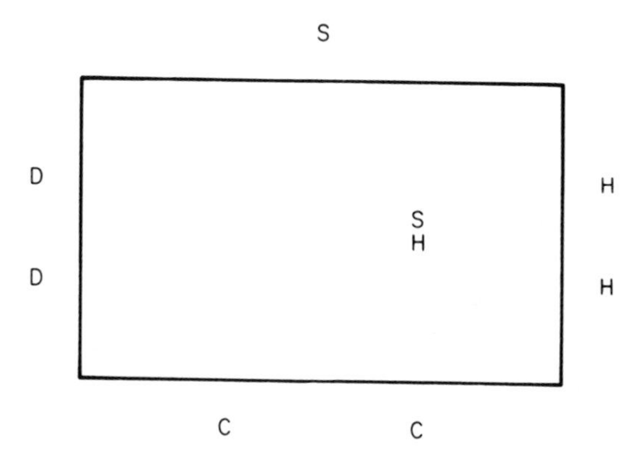

Fig. 18. The box is the playing area. The letters surrounding it designate cards in the possession of players.

Spades, who is thereby enabled to capture the pile. After the heart is killed, the situation looks like this (Figure 19)

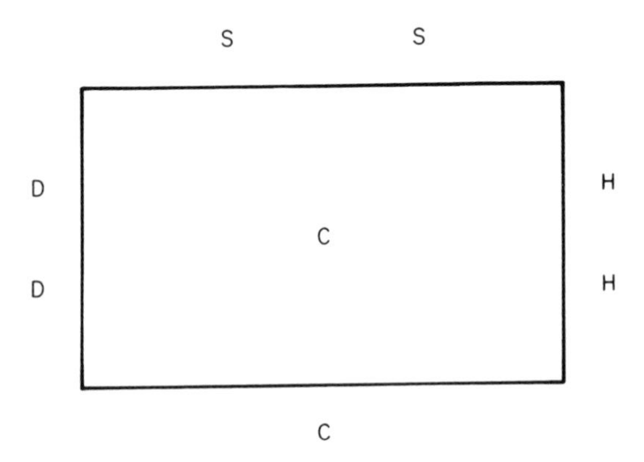

Fig. 19. The box is the playing area. The letters surrounding it designate cards in the possession of players.

It is now Spades' move. If the Black Coalition still persists, Spades, after having played a spade, will call on Diamonds, since, as has been noted, passing the move back and forth between two partners, when there is no hostile card to capture, only depletes their resources. Suppose Spades calls on Diamonds. Diamonds and Hearts now go through the same combination play as in the beginning of the game to capture either the club or the spade. (Had Spades played his card on clubs, they would have captured both cards.) If Diamonds made the capture, and if the card captured was the spade, we have the following configuration (Figure 20).

The move is again Diamonds'. He arranges to capture the club by putting a diamond on the club and calling on Hearts. Hearts captures his own card, which he must kill, and calls on Diamonds, who captures the club-diamond pile and kills the club. The next configuration is shown in Figure 21.

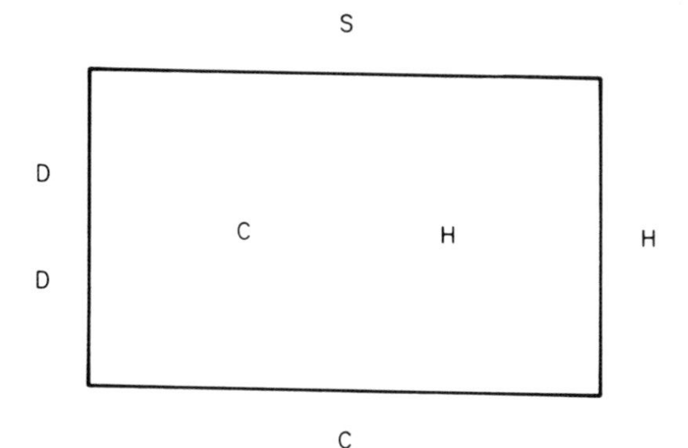

FIG. 20. The box is the playing area. The letters surrounding it designate cards in the possession of players.

The move is Diamonds'. He plays a diamond and calls on Spades. Neither Spades nor Clubs can now make a capture, because it takes two cards to make a capture. To prevent Diamonds from making another capture, Spades puts his card on the diamond. However, whatever Spades does, he will be out of the game when the move reverts back to him, unless he is "rescued" by his

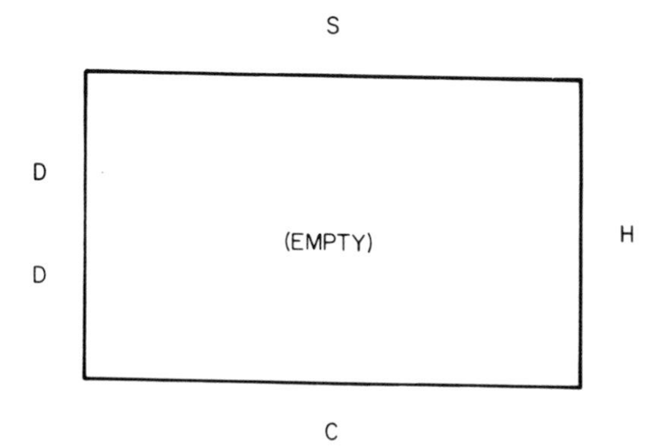

FIG. 21. The box is the playing area. The letters surrounding it designate cards in the possession of players.

partner (Rule 9) or by a member of the Red Coalition. Assume that the Red Coalition remains stable. Spades' partner Clubs can rescue him only by sacrificing himself. Therefore the outcome is a defeat of either Spades or of Clubs. Consequently it is of no advantage to Clubs (or to any other single player) to join in a coalition with the player who got the first move.

However, if *two* players join with the player who moves first, the three can defeat the fourth. Let us see how this can be done. Assume that Spades (to move first) is joined by Clubs and Diamonds against Hearts. Then Spades, after having played, calls on Hearts. Whether Hearts plays on the spade or separately, his card can be captured by a member of the coalition. Suppose he plays the heart separately (to prevent the spade from reverting back to the coalition) and calls on Spades (who has already spent one card). Spades plays on the heart and calls on Diamonds. Diamonds calls on Clubs, who plays separately and calls on Diamonds. Diamonds captures the pile and kills the heart. Diamonds now holds a spade in his position as a prisoner (Rule 8).

We shall suppose, for the moment, that the members of the coalition have agreed to "exchange prisoners" (Rule 8) and that they abide by the agreement. Consequently the spade is returned to Spades, and the configuration is as shown in Figure 22.

The move is now Diamonds'. Diamonds plays his card separately and calls on Hearts. Wherever Hearts plays his card, it can be captured; so he plays it separately. Suppose he calls on Spades. Spades plays on the heart and calls on Diamonds. Diamonds plays on the spade-heart pile and calls on Clubs. Now Clubs can play "without loss" by capturing his own card (which he kills), and calls on Diamonds. Diamonds captures the pile, kills the heart, and returns the spade to Spades. The configuration is shown in Figure 23.

Diamonds can now play "without loss" by capturing

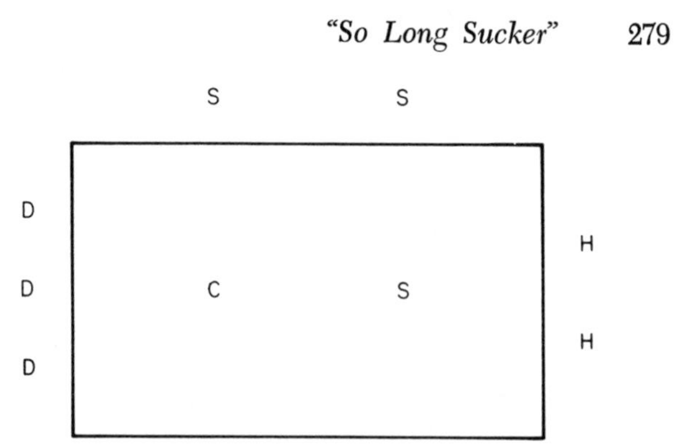

FIG. 22. The box is the playing area. The letters surrounding it designate cards in the possession of players.

his own cards. He then calls on Hearts. Again the heart can be captured by the same combination. Assuming Hearts had called on Spades, the result is shown in Figure 24.

Again the move is Diamonds'. He can now play his card and call on Hearts, who is thereby defeated, unless someone comes to his rescue. Assuming that no one will,

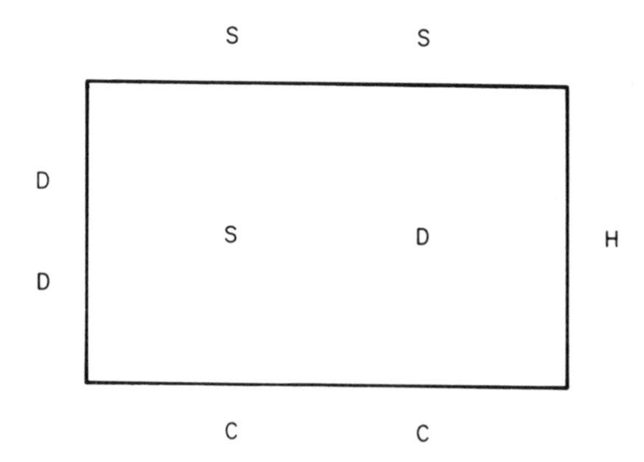

FIG. 23. The box is the playing area. The letters surrounding it designate cards in the possession of players.

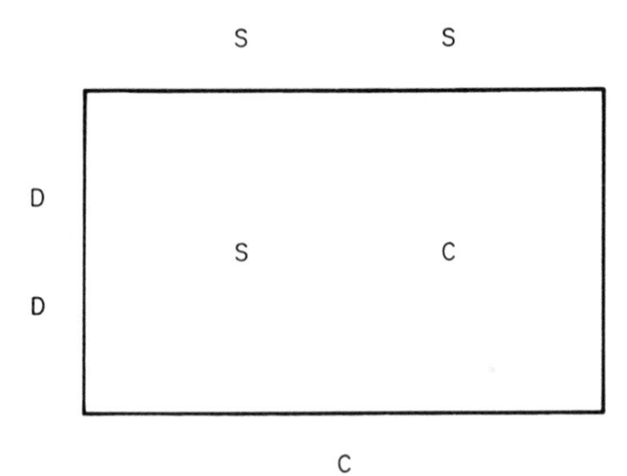

FIG. 24. The box is the playing area. The letters surrounding it designate cards in the possession of players.

what shall Diamonds do? He has three choices of where to play his card: on the spade, on the club, or separately. Recall that after Hearts' defeat, the move reverts to Diamonds. Therefore, he can capture either of the cards by playing upon it and by calling on Hearts. Suppose he decides to capture the club. Then the configuration will be as shown in Figure 25.

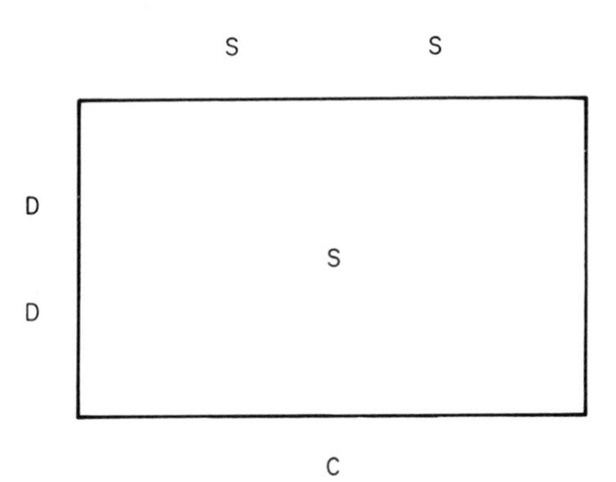

FIG. 25. The box is the playing area. The letters surrounding it designate cards in the possession of players.

Diamonds must move again. He and Spades can now put Clubs out of the game. However, it is Spades who must "make the kill" since, after Diamonds has played, he will not have two diamonds with which to make a capture. The play could go on as follows. Diamond plays separately and calls on Clubs. Clubs is defeated, but so is Diamonds after the next move. Thereby Spades remains the victor. On the other hand, Spades can be defeated, starting with the configuration in Figure 23, as follows. Diamonds plays on spades and calls on Spades. Wherever Spades plays, he must call on either Diamonds or Clubs, who will thereby cover his card and call upon him again. Spades must now play his last card and will be defeated when the move reverts to him next. The defeat of Spades requires the cooperation of Diamonds and Clubs. Clearly, however, after Spades' defeat, Diamonds and Clubs will be antagonists; hence their interests are in conflict *and this can be exploited by Spades.* Note that Spades and Clubs can also defeat Diamonds. After Diamonds has played, they cover his card and make him play again. Thereafter the player called upon again covers the diamond and calls upon Diamonds, whereby he is defeated.

So far, we have not considered the role of "prisoners." By Rule 6, the only cards with which a player can capture cards to retain in his possession are the cards of his own suit. He can also make captures by the use of prisoners; but the cards so captured must be turned over to the player designated by the capturing suit. Consequently, the possession of prisoners confers a bargaining power on the possessor. He can make captures for a partner in return for similar favors. Prisoners can be used also as cards to play to prevent being defeated, to rescue other players from defeat, and to give oneself the next move (which is of advantage if a pile can be captured thereby). All these possibilities add to one's bargaining power.

It must be kept in mind that in this wicked game no agreement is binding, but that it serves one's purpose to honor agreements in the beginning of the game in order to "build up credibility," which can be used later to doublecross one's erstwhile partners.

The full significance of bargaining (and double-crossing) comes into play if monetary payoffs are associated with the order in which the players are eliminated. Let us suppose that the first player to be defeated pays 20 units, the second pays 10 units, the next nothing, and the winner takes all (30 units). Then, before the first move is determined, the game is symmetric (cf. Chapter 15), and its characteristic function is as follows:

$$v(\bar{i}) = -20 \quad (i = 1, 2, 3, 4) \tag{18.1}$$

$$v(\overline{ij}) = 0 \tag{18.2}$$

$$v(\overline{ijk}) = 20 \tag{18.3}$$

$$v(N) = 0. \tag{18.4}$$

As soon as the first move is determined, the game ceases to be symmetric, if the first player and his single partner can be defeated in turn by the remaining two players (assuming, of course, that the coalitions, once formed, remain stable). Then, if player 1 has the first move, the characteristic function becomes:

$$v(\bar{1}) = -20 \tag{18.5}$$

$$v(\overline{12}) = v(\overline{13}) = v(\overline{14}) = -30 \tag{18.6}$$

$$v(\overline{34}) = v(\overline{24}) = v(\overline{23}) = 30 \tag{18.7}$$

$$v(\overline{ijk}) = 20 \tag{18.8}$$

$$v(N) = 0 \tag{18.9}$$

Nevertheless, as should be clear from our discussion, the characteristic function only scratches the surface of the game. It is based on the assumption that the coalitions once formed remain throughout the game. The *dynamics* of the game, however, are such that advantages accrue unequally to the members of a coalition as the game progresses. The game is sufficiently complex so that the relative values of these advantages are not immediately discernible. For example, in our hypothetical play, the Spades-Clubs-Diamonds coalition allowed Diamonds to capture the first pile. Thereby Diamonds acquired a spade as a prisoner. He had promised to return the prisoner to Spades. Is it worthwhile for him to break this promise? An extra card (of whatever suit) is always an advantage; but is this advantage worth the eventual loss of Spades as a partner? Note that certain rules of the game acquire significance in the context of establishing "credibility"; for example, the rule which allows a player to kill prisoners in his possession. This has no direct advantage, but it may be used as an act of "good faith" to give assurance to a prospective partner that the prisoners will not be used in plays against him when the two partners, having eliminated their opponents, face each other in the final showdown.

We gather what Martin Shubik meant when, in describing the game, he wrote, "The four inventors of the game still occasionally talk to each other."

19. The Behavioral Scientist's View

The task of the behavioral scientist is to describe behavior. In a scientific context, descriptions must be theoretical; that is, they must permit or suggest inferences stated in sufficiently general terms; they must bring out regularities or principles of behavior. Rationality is one such principle. This is not to suggest, of course, that rationality is to be sought in all or even in most instances of behavior. Attempts to do so amount to postulating an irrefutable hypothesis: whatever anyone does is rational from a certain point of view. This view leads to explaining behavior away, rather than to explaining it. Erasing the distinction between rational and non-rational behavior deprives the term "rationality" of informative content. It is perhaps unfortunate that "rationality" is not a neutral term. Denoting a form of behavior "rational" carries a connotation of approval. Conversely, to declare a form of behavior "non-rational" (not to say, "irrational") carries a pejorative connotation, which some behavioral scientists wish to avoid. It seems to them that the understanding of behavior involves an ability to put oneself into the state of mind of the subject, "to see things his way," and consequently to see the "rationale" which underlies any form of behavior.

A discussion on this level would lead us into questions concerning distinctions between "rationality" and "rationale," between "non-rational" and "irrational," etc. Whatever may be the intrinsic worth of such analysis, it would take us far afield from the present subject of discussion. We have already seen how "rationality" in the context of game theory must be further specified if the word is to have any theoretical leverage. One of the useful results of game-theoretic analysis is that it provides specific contexts in which the different meanings of rationality can be defined. Those definitions, then, can serve as base lines. Situations having a reasonable resemblance to games can be studied with reference to these base lines of rationality. It is the departures from these base lines, especially systematic pattern-forming departures, which call for explanations in psychological, sociological, perhaps even pathological terms.

To take an example, several definitions of "rationality" have been offered in the context of the N-person game. Some are so broad as to be useless without additional non-game-theoretic (i.e., non-strategic) specifications. Such is, for instance, the definition which leads to the Von Neumann-Morgenstern solution of the Three-person constant-sum game where every imputation is in some solution. Therefore, if "an outcome must be in a solution" is to be taken as a standard of rationality, the three players will have behaved "rationally" if only they did not throw away any money. A specification of a "standard of behavior" may single out a single solution from all the solutions of such a game.[47] But an imputation is, in general, contained in many solutions. Hence a single observation will not settle the question of whether the outcome was in the solution singled out. Only a whole series of observations on the same set of players (or on sets of players taken from a homogeneous" population) who we have some reason to believe are governed by the same "standards of behavior," can establish whether a solution (in the Von Neumann-Morgenstern sense) has been verified empirically.

The Shapley value and the Harsanyi bargaining model, on the other hand, do single out specific imputations for each game.[48] If the standard of rationality is defined as the adherence to the singled-out imputations, then systematic departures from it demand an explanation;

at which point the psychologist, the sociologist, or some sort of behavioral scientist should undertake the task.

Theories based on the bargaining set, the kernel, and ψ-stability lead to results intermediate between the Von Neumann-Morgenstern solutions and the Shapley-Harsanyi solutions. They single out classes of outcomes (usually narrower than the Von Neumann-Morgenstern solutions), each class associated with a particular coalition structure. They have nothing to say, however, about which coalition structures are likely to arise. These theories say, in effect, "*If* such and such coalition structure comes about, then such and such outcomes are 'rational.'"

If the behavioral scientist thinks about decision-making in conflict situations in the mode suggested by N-person game theory, he will focus on two fundamental questions: 1) Which coalitions are likely to form? 2) How will the members of a coalition apportion their joint payoff? It is noteworthy that N-person game theory has for the most part been concerned with the second of these questions, i.e., the question of apportionment. The one possible exception is implicit in the ψ-stability theory, in the sense that if some coalition structures cannot form any stable pairs with any imputations, it could be surmised that such coalition structures will not be frequently observed. In other theories coalition structure either does not appear in the "solutions" (as it does not either in the Von Neumann-Morgenstern, or the Shapley-Harsanyi models), or a coalition structure is simply assumed and the apportionment is derived with reference to it (as in the bargaining set and the kernel models).

The behavioral scientist, on the other hand, may be more fundamentally interested in the first of the two questions. In fact, seldom if ever can the behavioral scientist define payoffs in a conflict situation precisely enough to be realistically concerned with the question of their apportionment. Conflicts are all about us. People compete for affection, for power, for influence, for attention, etc., etc. Even if utilities could be assigned to "amounts" of these intangibles, it would be foolhardy, indeed, to assume a transferable utility in these contexts. Only money and possibly political patronage or the like can be cast in the role of a transferable utility.

In contrast, coalitions are often tangible and highly visible. Psychologists speak of "coalition structures" in families. The formal political process exhibits coalitions in specific situations (e.g., nominating conventions) with a precision which leaves no doubt. Large areas of international politics can likewise be described in terms of coalitions, forming and dissolving. Coalitions, then, offer a body of fairly clear data. The empirically minded behavioral scientist understandably is attracted to it. His task is to describe patterns of coalition formation systematically, so as to draw respectably general conclusions.

Analysis of the power of coalitions also constitutes a focus of investigation in N-person game theory. Accordingly, the behavioral scientist sometimes looks to N-person game theory for enlightenment. He does not find it there. Game theory is virtually silent on the question of which coalitions are likely to form; and even if this question is touched upon (as in the theory of ψ-stability), the answers, if any, are firmly linked to payoff configurations, which in most situations cannot be discerned.

Consequently the behavioral scientist spins his own theories of coalition formation, derived from what he knows—or thinks he knows—about human behavior. Such theories make contact with game theory only rarely and tangentially, partly because game theory lacks almost entirely the dynamic component, i.e., a model of the conflict *process*. We say "almost," because some of the ideas underlying some game-theoretical formulations do examine various situations in the process of their sequential formation. The game-theoretical emphasis, however, is entirely on the "stable" configurations, not on how they are arrived at. The main reason why game theory can say virtually nothing about the actual formation of coalitions in real life is because the predominating factors of real-life coalition formation are not "rational" in the game-theoretic sense. In real life, not what a coalition can accomplish, but the rewards of *being* in a particular coalition is what frequently holds coalitions together. To be sure, one can always make the argument that the rewards which accrue from simply belonging to a coalition are part of the coalition's payoff. However, as has already been pointed out, this sort of argument, being circular, will get us nowhere. It is like the argument that whatever anyone does appears rational if only one can understand the "rationale."

Nevertheless it seems worthwhile to examine some of the postulates behavioral scientists make regarding the formation of coalitions, and to relate them, if at all possible, to the ideas of N-person game theory.

The Minimum Resource Theory

This theory is embodied in the proposition that if a coalition is to accomplish something, then the coalition which commands the minimum amount of resources required to accomplish it will form.

The clearest interpretation of this principle is in terms of the simple N-person game (cf. Chapter 14). We assume that each potential member of a coalition is in possession of a certain amount of *resources*. A coalition of members whose pooled resources are sufficient to accomplish the task is a *winning coalition*, as defined in the theory of simple games. The minimum resource theory states that a coalition which possesses the minimal amount of resources needed to become a winning coalition will be the one most likely to form.

The rationale underlying this theory is a plausible one in the context of the simple game. Recall that, in the simple game, any winning coalition wins exactly the same amount of utility. If we assume that the resources are the "weights" of the players (cf. p. 217) and that the joint payoff is to be apportioned among the members of the winning coalition in proportion to the resources brought into the coalition by each of its members, it follows that no additional members will be welcomed into a *minimal* winning coalition. The larger coalition will not win more than the minimal coalition; it will simply have "more mouths to feed." Hence the members of a minimal winning coalition which has already formed will tend to become a "closed corporation."

It should be noted that this reasoning applies only to simple games. It need not apply, for example, to more general symmetric games in which the amount to be won by a coalition (the value of the game to it, as given by the characteristic function) depends on the number of members. There the pro-rata gain (cf. Chapter 15) is the important concept. To be sure, the minimum resource principle can be modified to take the pro-rata gain into account. It could read, for example, as follows: The coalition which will tend to form is that smallest coalition which can give its members the largest pro-rata gain. However, the rationale, plausible in the context of the simple game, is much less plausible in the present context. For, consider two coalitions S and T, such that $s = |S| < |T| = t$, both coalitions affording the largest pro-rata gain to its members. The minimum resource principle applied in this context would predict that S would be more likely to form than T. However, it can no longer be argued that the admission of the members of the set $T - S$ into S, so as to form T, means a loss to the members of S, since their pro-rata gain in T will be the same as in S. It can, perhaps, be argued that once S has formed there is no need to recruit more members into it; and so the principle of economizing effort leads to the conjecture that S rather than T will form.

We have seen, however, that in a game of this sort a finite symmetric solution (cf. p. 226) apportions the largest payoffs among the members of the larger, not of the smaller coalition. Roughly speaking, the resulting imputation is "more stable" because fewer members are disgruntled; and we have seen how a sufficient number of disgruntled players (not necessarily a majority) can "wreck" any imputation by going off to form a coalition of their own. Here, then, an opposite principle seems to be operating: "Include as many as possible in a coalition, if this can be done without loss to the members." In some political contexts, this, too, is a sound principle.

Clearly, the contexts in which the two principles apply are different; so that there is no real contradiction among them. The distinctions between contexts are, however, often obscured when purely verbal statements of principles are taken seriously. Thus the principle of minimum resources can be made even more vague if it is interpreted to mean that "the weak unite against the strong." In some contexts, this can be argued. Assume, for example, that at a nominating convention, one of three candidates (A) controls 48 percent of the votes, another (B) 30 percent, and a third (C) 22 percent. Who will be nominated? If we apply the minimum resources principle, we expect that B and C will form a coalition against A and, consequently, one of the two weaker candidates will be nominated. Here the weak are expected to combine against the strong.

However, if we accept this principle without distinguishing carefully the senses in which a player can be "strong" or "weak," we may get into serious difficulties. Recall the experiments reported by Maschler (cf. Chapter 17), in which the strongest of three players was the one who was a member of the two coalitions commanding the two largest values among the values to two-person coalitions, and the weakest player was the one who was a member of the two two-person coalitions which commanded the two smallest values. If we had predicted the coalitions of the two weaker players against the stronger in that context, we would have been right in only three cases out of sixteen.

Nevertheless the principle that "the weak unite against the strong" may be defended on psychological grounds and may indeed be observed here and there. The corroboration of this principle in specific instances should not, of course, be interpreted as its vindication. Rather, the particular aspects of the specific situations ought to be examined to see whether they provide a rational basis for this phenomenon; and, if not, some rationale specific to the situation rather than a generally valid principle should be invoked in constructing an explanation.

The Minimum Power Theory

William Gamson[49] lists a "Minimum Power Theory" as another principle of coalition formation and cites the Shapley-Harsanyi solution as a foundation of this principle. According to this theory, the coalition that will form (in a simple game) is the winning coalition which has the minimal power (e.g., joint Shapley value) among all the winning coalitions. Naturally, in a simple majority game (one man, one vote), the two theories coincide. In a weighted majority game, however, it is easy to find examples where the predictions of the two theories differ.

Consider a Four-person weighted majority game with the players having 10, 5, 3, and 3 votes respectively, where a two-thirds majority is required to win. The winning coalitions are ($\overline{12}$), ($\overline{123}$), ($\overline{124}$), ($\overline{134}$), and ($\overline{1234}$). Player 1, who has a veto, has 7/12 of the power.

Player 2 has 3/12, and player 3 and 4 have 1/12 each. It follows that coalition $(\overline{134})$ is the winning coalition with minimal power. However, coalition $(\overline{12})$ is the winning coalition with minimal resources. Here the two theories make different predictions, and in principle they could be put to a competitive test.

The Anti-Competitive "Theory"

This "theory," also cited by Gamson, derives from purely sociopsychological considerations. It emphasizes the conflict-reducing motivations of the members of a group who find themselves in a conflict situation. We put this "theory" in quotation marks because the level on which these matters are discussed is furthest removed from the level of theory formulation, as theory is understood in a scientific context. This is by no means to suggest that the sociopsychological factors are of little importance in understanding the behavior of people in conflict. On the contrary, they are often of overriding importance, and their exclusions from game theory seriously weakens game theory as a tool of social science. Rather, the quotation marks are meant to indicate that the usual discussions of these matters are formulated in a way which leaves them with little theoretical leverage. They "call to mind" certain aspects of conflict, but do not define them in a way that makes it possible to include them in a well-formulated theory. Specifically, the anti-competitive "theory" maintains that people do not only pursue their individual interests, that they also want to "preserve the coherence of the group." Such conjectures are plausible; but what follows from them? What will people specifically do in a game-like situation? Will they form a grand coalition? If they do, will they divide the joint payoff equally? Neither is a safe prediction.

Nevertheless certain aspects of the anti-competitive theory can be translated into game-theoretic terms. One could conjecture that in certain circumstances the *largest* coalition, consistent with certain goals of the individual players, might be likely to form, which pits the Anti-Competitive theory against the Minimal Power Theory.[50] Or, one could conjecture that the final apportionment of payoffs will depart from the apportionment suggested by the relative power positions of the players in the direction of a more equal division. The theory may also be interpreted as saying that "conflicts will be minimized within coalitions," where "conflict" in this context no longer means the original conflict of interest, as defined in the game, but other sources of conflict, rooted in the personal ties of the players. In other words, this may be just a way of saying that birds of a feather flock together. To put it somewhat more concretely: other things being equal, people who are "most congenial with each other" will form coalitions. Or putting it more suggestively, one could guess that other payoffs accrue to coalitions besides those specified by the game. But what are these payoffs? Are they comparable with the payoffs of the game? Are they in transferable utilities? Unless these questions are answered, the game-theoretician's argument, to the effect that *all* utility accruing to a player is represented in his payoff, is no more than a formalistic argument. We have already argued that this is so (cf. p. 284).

It appears, then, that the sociopsychological factors can at best be viewed as perturbing influences in game theory. For example, if certain solutions of symmetric games imply that the largest coalition consistent with the pro-rata gain will form (cf. p. 226), then, if the observed coalitions are *still* larger, one can ascribe the discrepancy to sociopsychological factors, brought out by the anti-competitive theory. If the apportionment of payoffs is more egalitarian than predicted by the Shapley value solution, the same conjecture can be made. If the players are in a position to employ threat strategies in their bargaining (as in the case in Harsanyi's formulation), but the final solution is more in accord with the Shapley value than with Harsanyi's bargaining model (where the two solutions are different), then, perhaps, a concrete meaning can be ascribed to the statement that "coalitions will form . . . along the path of least resistance in bargaining," in addition to the meaning ascribed to it by Gamson, namely that coalitions will tend to form among players with approximately equal resources.

Experimental Evidence

In Chapter 17 we examined N-person game experiments aimed at putting to a test hypotheses derived from formal game-theoretic concepts, e.g., Shapley value, ψ-stability, bargaining set, etc. Here we shall examine some results of experiments designed by behavioral scientists (rather than game-theoreticians) with a view of putting to a test the theories mentioned in this chapter. There is an overlap, of course, between these theories and those derived from game-theoretic concepts; but, as I hope has become clear, the emphasis is different. The point of departure of the behavioral scientist is the question "How do people behave in conflict situations?" He attempts to answer the question by designing a game to simulate the situation. The point of departure of the experimenting game theoretician is "How is the strategic structure of a game reflected in the way people play it?" He attempts to answer it by designing games in which some crucial structural feature is varied, and by then observing differences in behavior.

If the theories mentioned above are to be tested, they should be pitted against each other. In the present case, it is possible to pit the Minimum Resources Theory against the Minimum Power Theory by designing games in which the pivotal powers of all players are equal, but not the resources. A stronger experiment would be one where the predictions of the two theories are incompatible; but, to my knowledge, this has not been done.

Consider a majority game in which player 1 has three votes, while player 2 and 3 have two votes each. Clearly, each player has the same pivotal power in this game, whose characteristic function is $v(i) = 0$, $v(\overline{ij}) = v(\overline{123}) = 1$. Player 1, however, has "greater resources." The "resources" will enter the picture only if the players assume that they are relevant in apportioning the prize among the winners. The characteristic function, however, says nothing about this. The Shapley value, based on pivotal power alone, accords equal portions to all three players. If a pair coalition forms, the Shapley value solution does not apply. As for the core, the game hasn't any. The Von Neumann-Morgenstern solution says that *if* one player gets $\frac{1}{2}$, then another must get nothing, but says nothing about which coalition, if any, will form, nor anything about how the other two will split, if the first player gets something other than $\frac{1}{2}$. From the characteristic function it is apparent that the "resources" are irrelevant to the outcome of the game. Hence any results which are most conveniently explained by the Minimum Resources Theory corroborate the theory, which, incidentally, in this example makes a prediction at variance with *all* game-theoretical predictions!

W. Edgar Vinacke used female subjects in experimental Three-person games in which the pivotal powers, but not the "resources" of the subjects, were equal. His results are shown in Table 17.[51]

In Games 1 and 3, the Minimum Resources hypothesis seems to be corroborated. Although each pair coalition has the same power as a minimal winning coalition, the most frequently observed coalitions are those between the two players with the smaller resources. In Game 2, the results are inconclusive, since two of the three possible pair coalitions are consistent with the theory, and, in fact, about two-thirds of the observed coalition are

TABLE 17

Distribution of resources				Predicted coalition (numbers indicate players)	Observed coalitions		
					Predicted	Other	None
Player	1	2	3				
Game 1	3	2	2	$(\overline{23})$	64	25	1
Game 2	1	2	2	$(\overline{12})$ or $(\overline{13})$	64	15	11
Game 3	4	3	2	$(\overline{23})$	59	29	2

in those categories. However, it may be noteworthy that in this case no coalition was formed in 11 cases. This may indicate that in the absence of a unique minimal resource coalition the choice becomes harder. Evidently there is reason to believe that "resources" influence coalition formation even if they play no part in the theoretical outcome of the game.

Gamson conducted experiments on Five-person games in which resources played a more direct role in the rules of the game. Specifically, the players were told to imagine that they were chairmen of delegations at a political convention. Each controlled a given number of votes. The task was to form a majority coalition with the view of winning political patronage in the form of jobs. A coalition was considered to have formed if a set of members controlling a majority have declared themselves a coalition and have agreed on how to divide up the spoils. The resources (votes controlled) of the respective players were 17, 17, 17, 25, and 25. Thus the game was actually a symmetric weighted majority game, every three-person coalition being a minimal winning coalition.

Gamson observes that the coalition $(\overline{123})$ was formed in 33 percent of the cases, whereas the a priori probability of this three-person coalition is only

$$\frac{3!2!}{5!} = \frac{12}{120} = \frac{1}{10}. \qquad (19.1)$$

This result also seems to corroborate the Minimum

Resources Theory. B. Lieberman[52] performed experiments using a Three-person game, in which coalition $(\overline{12})$ received 10 cents from player 3, coalition $(\overline{13})$ received 8 cents from player 2, and coalition $(\overline{23})$ received 6 cents from player 1. The apparently weakest coalition $(\overline{23})$ was observed to form most frequently. Lieberman comments ". . . if [player 1] was seen as the strongest and most exploitative individual, this may have led [player 2 and 3] to unite against him . . . ." The result seems at first sight to be at loggerheads with the results obtained by Maschler (cf. p. 267), where the two "strongest" players united most frequently. (Here it is instructive to postpone a controversy concerning the validity of the "union of the weakest" hypothesis until in the concluding chapter the concept of "strong player" and "weak player" are clarified in the various contexts.)

Observe that Lieberman's game is a zerosum game. Consequently it is strategically equivalent to the game with characteristic function $v(\overline{i}) = 0$, $v(\overline{ij}) = 1$, $v(\overline{123}) = 1$ $(i = 1, 2, 3)$, the one and only Three-person essential constant-sum game there is (cf. p. 85). In this game, properly normalized, the bargaining positions of all three players are equal. They *appear* unequal because of the way the numerical values of the payoffs were chosen. Now, it may very well be the case that the players who *see* themselves as the weakest pair tend to unite in a coalition for that reason, which is a corroboration of the "union of the weakest" hypothesis. But the Anti-Competitive hypothesis has an equal claim. Observe that player 1 stands *least to lose* if he is left out of the coalition (namely 6 cents), whereas player 2 stands to lose 8 cents, and player 3, 10 cents. The greater frequency of the $(\overline{23})$ coalition may well be a reflection of a tendency to minimize the loss to any one player (as well as a reflection of "resentment" against the most privileged player).

Maschler's games, on the other hand, are all non-constant-sum games. Moreover, they are normalized in such a way that any player left out gets the same payoff (0). Consequently the reluctance to minimize the loss to the "victimized" player does not operate, while the tendency for the "strongest" coalition to form (in the sense of the largest pro-rata gain) does.

Examining the patterns of payoff distributions in games where the players have equal pivotal powers but unequal resources, we find that the Minimum Resource Theory tends to be corroborated qualitatively but not strictly quantitatively. That is, the coalition members with larger resources tend to get larger shares of the prize but less than proportionately larger. Again an "equalizing" bias seems to be operating.

Real Life Data

The behavioral scientist who looks at "real life" for clues on where the insights from game-theoretical analysis can be applied, faces a much more difficult problem than the experimenter. Instead of setting up conditions relevant to such analysis, the behavioral scientist must interpret events as they occur. As we have said, coalitions are the clearest observables in real life conflicts which relate to the essential ideas of N-person game theory; and of all types of coalitions, those which manifest themselves in well defined acts (e.g., forming blocs of votes) are the clearest.

William Riker[53] applied what he calls the principle of minimal size to the presidential election of 1824.

The standings of the four candidates in the electoral college were as follows:

Andrew Jackson	carried	11	states	with	99	electoral	votes
John Quincy Adams	"	7	"	"	84	"	"
William H. Crawford	"	3	"	"	41	"	"
Henry Clay	"	3	"	"	37	"	"

No candidate having obtained a majority, the election reverted to the House. In such cases the House votes by states, the single vote of each state going to the candidate favored by a plurality of the representatives.

Therefore the situation became the following:

Jackson	11	votes (states)
Adams	7	" "
Crawford	3	" "
Clay	3	" "

Under the Twelfth Amendment, however, a maximum of three candidates could be considered. Clay, who received the fewest electoral votes, had to withdraw. To whom will Clay's votes go?

At this point, horse trading begins. Riker notes that with the above distribution of votes, Adams, Crawford, and Clay are potential "members of a uniquely preferred minimal coalition" and therefore Jackson is strategically weak. It is not clear, however, whether in invoking the principle of minimal size Riker has in mind the principle of minimal resources or the principle of minimum power. If the game is a weighted majority game, then, using the Shapley value index of power, we arrive at the result that Jackson holds 50 percent of the total, while Adams, Crawford, and Clay share the other 50 percent equally (in spite of the disparity in their resources!). Be that as it may, the Adams-Crawford-Clay coalition is a winning coalition which has both minimal resources and minimal power. If we agree that this puts Jackson in a strategically unfavorable position, we may expect that, in the horse trading, Jackson is going to lose support (as people switch to the side expected to win). And he does. By the time the House was ready to vote, the score stood as follows:

Adams	11
Jackson	7
Crawford	3
Clay	3

Now the Clay-Adams coalition is minimal-winning, both with regard to resources and power. Clay throws his support to Adams; Adams becomes President, and Clay the secretary of state, which in those days was considered the most advantageous springboard to the presidency.

How seriously are we to take these calculations? Are not the conventional political explanations of the Clay-Adams alliance more convincing? I do not know. If we could find situations where the predictions of "conventional" political theory and of the behavioral scientist's interpretation of N-person game theory seem incompatible, we could pit one against the other. In the above instance they are compatible, and the question remains open.

20. *Concluding Remarks*

The principal lesson to be drawn from a study of game theory is, in my opinion, a realization of how much must be clarified before one can even raise the question "How do I act in this conflict situation?"—a question which is constantly put by people who have the power of life and death over the inhabitants of this planet.

This lesson is especially pertinent in our age, dominated by decision makers who owe their careers in considerable measure to "management ability." In public life, especially in the United States (and now increasingly in the USSR), "management" is understood in two contexts, economic and administrative. The management of a business enterprise involves choices of actions based on estimates of possible outcomes and of their utilities. The utilities are often estimated in terms of further probable consequences and ultimately in terms of explicitly or implicitly assumed goals. The difference between the management problems of a "capitalist" and a "socialist" enterprise is that in the former competitive goals often play a major role (except in monopoly). Whatever be the social consequences of the presence or absence of competitive goals, the techniques of economic management in all technically advanced societies are strikingly similar. The "manager" is essentially playing a game against nature. His skill depends very largely on correct guesses of probable outcomes, which, in turn, depend on his own actions and on other circumstances that he does not control but whose "probabilities" he roughly estimates.

The administrative problems of management are those

of organization: the apportionment of tasks, chains of communication, and hierarchies of authority among people and agencies with the view of implementing most efficiently the decisions taken by management.

The concept of conflict is most familiar to the management archetype in two principal contexts, business competition and war. He may therefore become aware that the One-person game paradigm of decision making is not an adequate framework in which to formulate his problems. The war-maker is especially cognizant of this; in fact, he has been so since the beginnings of war. Consequently the art of strategy is recognized as an important component of military proficiency (and to some degree of business acumen). Therefore the manager (be he a business leader or a war leader) is likely to be receptive to the idea (if he is interested in ideas) that a groundwork of a *science* of strategy is being laid by the theory of games, to supplement the *art* of strategy, where proficiency depends largely on individual talent. This is especially true in a technically advanced business society where both machines and organizations are designed to operate with interchangeable parts, with a view of diminishing the dependence upon either superior craftsmanship or superior leadership.

Now, we have warned the reader in Chapter 1 that game theory is *not* a science of strategy, not even an abstract groundwork for a science of strategy; and I hope this has become clear from the contents of this book. This is the negative aspect of the first lesson to be drawn from game theory. The positive aspect is the revelation of the enormous complexity of the problem of *formulating* the logical structure of a conflict situation. As we pass to the higher realms of conflict theory (i.e., the non-constant-sum games and especially the N-person games), the problem of formulation lays more and more claim to our attention, and as we pursue the analysis, we arrive at conclusions that have no bearing on the original problem which may have instigated the analysis: "What is the best strategy (for me) in a given conflict situation?" We find in the theory of the N-person game a plethora of problems involving the power relations among coalitions, in which the strategies of the original game have been completely abstracted from. If we wish to solve *these* problems (which must be solved if the original problem is to be *clarified,* let alone solved), we must forget the strategies of the original game and concentrate on the power relations among all the possible coalitions.

This in itself may not dispel the interest of the decision maker. The modern decision maker, especially the political leader, may be well aware of the importance of coalition-forming strategies. He may therefore be willing to forget action strategies for a while and turn to N-person game theory, hoping to find therein some guidance on effective coalition-forming strategies. As we have seen, he will find very little that will be of use to him. The closest thing to the sort of answers he may want is the theory of pressure groups (Chapter 16). There the objects of investigation are the strategies available to a coalition to break up into "pressure groups" in situations where a commonly accepted standard of fairness determines the disbursement of a cooperatively achieved payoff among the pressure groups.

The main difficulty encountered by the decision maker searching game theory for guide lines in the design of strategies (whether of action or of coalition formation) is in the ambiguity of the identity of the actor. The modern decision maker tends to take the identity of the actor for granted. The actors are he himself and his opposite numbers, or else the organization he serves and its opposite numbers. It is the firm or the department or the institution or the nation, whose "interests" are either given or are to be discerned in some manner and, once discerned, are pursued. The decision maker is therefore likely to look at game theory for indications on how an individual "rational actor" (however he is identified) should act.

We have seen, however, that N-person game theory would be crippled without the distinction between individual rationality and group rationality. Moreover, "group rationality" in that context does not refer to the collective rationality of the *in-group,* which the administrator and the military leader may understand very well, but to the collective rationality of *subsets of players,* who have partially conflicting interests that *cannot* be "forgotten," as they must be if an organization or an institution is to behave as a single actor.

Moreover, these subsets are not fixed but fluid; and their overlapping, partially coincident, partially conflicting interests must be kept in mind simultaneously (in counterpoint). This is not simply a problem of handling a large complex of "factors" (such as must be taken into account in logistics and in cost-accounting in business and in war). It is a problem involving continual shifts of focus to follow the *changes of identity of the actors,* something basically foreign to the mode of thinking of those who engage in strategic conflicts.[54]

Another example of a time honored (or dishonored?) concept which could be subjected to merciless analysis with the help of rigor encouraged by game-theoretic formulations (with due regard for the identity of the actors) is the so-called "balance of power" in international relations.

On previous occasions, writing about the conduct of international relations, I have repeatedly criticized the "strategic mode of thought." This has been interpreted at times as a criticism directed at applications (or misapplications) of game theory in designing foreign policy. Vigorous rebuttals were given to these views (attributed to me, perhaps not unreasonably considering the way some of my arguments were exemplified) by members of the "strategic community." These rebuttals first called attention to the fact that little or no formal game theory was used in the design of foreign policy (which is, of course, true); second, that all rational thinking is "strategic," even if the humanitarian values are included in the design of policies. What I was actually criticizing under the designation of "strategic mode of thought" is the fixed identification of the actor habitual to this manner of thinking. Once the identity of the actor is fixed, the question "How shall he act in a conflict situation?" seems to make perfect sense. But it is this very question which cannot be unambiguously answered by game theory in any but the most primitive of conflict situations, such as the Two-person constant-sum game. In more complex situations, ambiguities and paradoxes arise which cannot be removed as long as the inquirer remains an actor *inside the game*, as it were. Definitive answers can be given only by the *outside* observer, who addresses the whole set of players as "the actor." This outside observer (e.g., Shapley or Harsanyi) has taken into account both the interests of the individual players and the interests of the several subsets as potential coalitions, as well as the strategic options open to all of these "actors." On the basis of all these considerations he proposes a disbursement of payoffs among the players, i.e., a *settlement*. In this context, game theory emerges not as a theory of strategic conflict but rather as a theory of conflict resolution.

In a way, such settlements can be viewed as stable configurations based on some sort of balance of power principle and, as such, may have a certain appeal to those members of the strategic community who also prefer to think in these terms. On this, two observations are in order. First, the so-called definitive solutions of an N-person game (e.g., Shapley's, Harsanyi's) are Pareto-optimal solutions. The grand coalition cannot do better jointly (in Shapley's) nor better for every individual player (in Harsanyi's) than by adopting the proposed solution.

Second, the "balance of power" is not preserved by

power alone. Thus the solutions differ radically from the "balance of power" solution of the Two-person constant-sum game, in which neither player *can* get a larger payoff than the value of the game to him. In mixed-motive or N-person conflict, the "balance of power" is notoriously unstable. Its preservation depends on a "higher rationality" of the players, which takes into account not only the immediate gains of sporadically formed coalitions but also the ultimate consequences. In effect, this balance of power can be preserved only by invoking the Kantian categorical imperative: act in a way such that if every one acted in this manner it would be in your interest. In the context of the N-person game, *"you" means not only the individual player but also every potential coalition of players.*

It serves little purpose to argue that the "balance of power" principle, practised by the European states in the pre-Clausewitzian era (1648–1792), can be resurrected in the present super-power era. What may have been Pareto-optimal for the assorted princes and princelings was hardly likely to have been Pareto-optimal for their populations. But the populations of eighteenth-century Europe had little or nothing to say about the miserable little dynastic games which led to the "cabinet wars" of those days. The actors of today are not those of the cabinet wars. No balance of power based on "deterrence," the favorite principle invoked by the "reasonable" strategists, is anywhere near Pareto-optimality, if the needs, the aspirations, the very survival potential of the populations are taken as the payoffs of the "game." Finally, the tolerance to perturbations in the "balance of power" in the nuclear age is not comparable to its counterpart in the days of marches, counter-marches, sieges, and encampments. The only realistic "settlement" in our age can be an equitable settlement, not a power-enforced settlement, simply because the number and diversity of potential "actors" [55] transcends the possibility of a genuine balance of power. For this reason the models of a "stable world" proposed by the modern enthusiasts of the balance of power principle bear little or no relation to the insights derived from "higher" game theory.

Another area into which game theory has brought much-needed clarification is that of utility theory.

The analysis of multi-sided conflict of interest (formalized as an N-person game) proceeds along different lines depending on whether or not the actors can form coalitions and, if so, depending on whether or not their utilities can be compared and transferred. In Von Neumann's and Morgenstern's exposition, comparable and transferable utilities were introduced in passing from Two-person to N-person games. Indications are that this was done for methodological reasons. In particular, the characteristic function was defined in terms of sums of utilities of different players. Attempts to develop an N-person game theory without assuming comparable, transferable utilities (cf. Chapter 11) seem to have been stimulated by a train of thought in a school of economics which rejects the idea of comparable utilities. Certain problems of theoretical economics previously treated in terms of a utility function were shown to be tractable without this notion, and consequently the very concept of utility was dropped, presumably in the interest of parsimony. It is clear, however, that aside from the classical market, few social institutions could function without tacit comparisons of utilities of different actors. In particular, the concept of social justice could not be defined without recourse to such comparisons. It seems, therefore, that the question of admitting or excluding the notion of comparable utilities is not merely a methodological question. It may have risen in a purely methodological context, but it reflects important differences of social philosophy. Here we have an illuminating example of how game theory brings certain fundamental questions to the center of attention and puts them under the intense light of rigorous analysis.

Finally, an important lesson to be drawn from the analysis of the essential ideas of N-person game theory relates to the methodology of experimental behavioral science. The principal task of the behavioral scientist is to describe systematically, and to explain in terms of general principles, patterns of observed behavior. Increasingly the behavioral scientist has turned to the laboratory, seeking answers to specific questions with reference to behavior under controlled conditions. Game theory has given a stimulus to these inquiries by providing a rigorous framework for constructing theoretical models. Thus a behavioral scientist who wishes to put some conflict theory to a test can simulate a species of conflict as a game and translate the predictions of his theory into predictions concerning the outcomes of the game. In doing this, the behavioral scientist may discover that the terms in which his theory is formulated are much too vague to be translated unambiguously into game-theoretical concepts. He may then postulate some intuitively plausible principle like "maximization of gain" or "minimization of effort" or "competitive advantage" or "equitable resolution of conflict." What game-theoretic analysis reveals is that the apparent clarity of these terms is deceptive. Nor do attempts to define these terms "operationally" always help. Definitions may *look* operational without being so in relation to what is clearly observable.

The behavioral scientist will be well advised to acquaint himself with the intricacies of game-theoretical analysis beyond the elementary concepts of strategy, payoff, coalition, etc. The value of doing so is not in that game-theoretic analysis is a replica of how people actually analyze conflict situations. In fact, people in conflict do not often engage in analysis at all. The value of game-theoretic analysis is that it suggests experiments or observations which would not have ordinarily occurred to the experimenter.

Such experiments can then be performed with the specific purpose of testing the game-theoretic models (instead of the usual behavioral hypotheses). The former have the advantage of at least logical precision, although much is left to be desired in empirical precision (e.g., the determination of subjective utilities). The logical precision of game-theoretical formulations allows the experimenter to vary his independent variables, especially the strategic structure of the game, while being in complete control of those variables. The independent variables usually considered or manipulated by behavioral scientists (e.g., personality of the subjects, experimental environment) for the most part lack this precision. Thus, even though the *dependent* variables of behavioral experiments (i.e., the observables) can be quantified and "hardened," the rigorization remains one-sided. The typology of N-person games (constant-sum, non-constant-sum, with and without cores, symmetric, simple, quota, etc., etc.) provides a rich repertoire of situations wherein the differences are often not perceptible to the naked eye, but which lead to often quite disparate "solutions," of which game theory also provides a plethora.

As an example, consider the concept of the core. The core is a plausible candidate for a range of "stable" imputations of an N-person game. These imputations keep all the subsets of the players "happy," or at least deprive them of a motivation to leave the grand coalition. Constant-sum games, however, have no cores (in itself an important insight into the nature of a generalized pure conflict game). Thus the idea occurs to create "artificial cores" in such games by imposing restrictions on re-alignments of players in coalitions. That is, when a certain amount of "social friction" is introduced, certain configurations of coalitions and payoffs become stable which had not been stable in the completely fluid situations (another idea suggestive of real life as well as of certain physical systems). Thus a promising area of experimentation is suggested by a particular development of N-person game theory.

It seems advisable to be guided in the design of "conflict experiments" by this exceedingly rich theory. For this purpose it is necessary to disregard for the moment the question of what sort of conflicts all these games "represent." Suffice it to say that they represent a whole universe, strictly taxonomized into phyla, classes, genera, and species. Is this classification reflected in any way in a taxonomy of real behavior? If so, can it be that, even though intricate game-theoretic analysis is beyond the conceptual repertoire of the ordinary subject, nevertheless the subjects' intuitions somehow parallel the analysis? If not, wherein lie the differences between the taxonomy of games and the patterns of behavior of players?

It seems to me that answers to these questions will *naturally* transform themselves into hypotheses related to behavior in conflict situations, and that these hypotheses, being the outcomes of an interaction between a rich theory and experiments suggested by the theory, will lead to a more systematic development in the study of conflict.

It may even happen (although this is not likely) that no relations at all will be observed between the strategic structure of games and the behavior of subjects engaged in them. If so, we shall be justified in dismissing "rationality" altogether in accounting for conflict behavior. More likely, "rationality" (of various levels, of which there are many, as we have seen) will play a greater or a lesser part in games with different strategic structures; and this may lead us to important questions concerning how rationality of various levels impinges or fails to impinge on human conflict.

Notes

1. To every positive real number there corresponds a unique real number as its logarithm. Defined in this way, logarithms of negative numbers do not exist. To avoid this restriction, the definition of logarithms is extended to complex numbers. In this domain, to every complex number, except zero, there corresponds an infinite set of complex numbers as logarithms. Any two of this set differ by a multiple of $2\pi\sqrt{-1}$.

2. This logical restriction underlies Bertrand Russell's theory of types and other concepts of modern mathematical logic.

3. A strategy is essentially a plan chosen by a player in which the player's choice among available alternatives is indicated in every situation that can arise during a play of the game. For a fuller discussion of this concept see the author's *Two-person Game Theory,* Chapter 3.

4. Note, however, that "brother of" is not a transitive relation since "x is the brother of y" and "y is the brother of z" do not imply "x is the brother of z" if x and z are the same person.

5. It could be maintained, of course, that to some extent chance governs the course of the chess player's deliberations (via unknown, hence "random" fluctuations of the state of his nervous system). These events, however, do not enter the definition of the rules of chess, and therefore play no part in the game-theoretic analysis of this game. On the other hand, the shuffling of cards, throws of dice, etc., do enter the rules of some games of strategy. In such games Chance is considered to be taking part in the game.

6. In a simple version of the game of Nim, fifteen sticks are arranged in three rows of three, five and seven sticks respectively. A move consists of picking up any number of sticks from any one row. The players move alternately. The player who picks up the last stick loses. In this game the player who moves first must always win if he plays correctly.

7. If a game has a termination rule which insures a finite number of moves, the number of pure strategies available to each player is finite, although in general enormous. The mixed strategies, on the other hand, constitute a continuum with as

many dimensions as there are strategies. The outcomes of mixed strategy choices are understood as *expected outcomes,* that is, averages over all possible outcomes weighted by the corresponding joint probabilities of strategy choices.

A concept related to mixed strategy is the *behavioral strategy,* in which a player indicates the probability with which he will choose each of the available alternatives in every situation. If a game satisfies certain conditions, it can be shown that a player does not suffer a loss of strategic versatility if he confines himself to behavioral strategies (which are conceptually simpler than mixed strategies). In some games (of which Bridge is a famous example) this is not the case if each pair of partners is considered as a single player. For further discussion of this topic, see Luce and Raiffa, *Games and Decisions,* Chapter 7, and G. L. Thompson, "Signaling strategies in n-person games."

8. *Games of complete opposition* are those in which the preference orders of the two players with respect to the set of outcomes are exactly opposite. These include other than constant-sum games. In fact, if the payoffs of one player of a constant-sum game are multiplied by a positive constant, the game is, in general, no longer constant-sum, but it is still a game of complete opposition. In game theory, the concept "non-constant-sum game" usually refers to a game which is not a game of complete opposition.

9. Note that reference to "joint payoff" implies that payoffs of different players can be added. This assumption is usually made in N-person game theory but not in Two-person theory.

10. Formally, a weak ordering among imputations can be established by defining all the imputations of a set equivalent if they are members of an intransitive cycle (as well as pairs that dominate or do not dominate each other). However, such a definition of the order relation would be useless in most cases: too many of the imputations would thereby become "equivalent."

11. However, in games where they exist, cores form an important anchor point of analysis. For discussions of this topic, see M. Shubik, "Edgeworth market games" and Y. Kannai, "Continuity properties of the core of a market game."

Whenever an N-person game has both a Von Neumann-Morgenstern solution and a core, the set of imputations in the core must be contained in the set of imputations in the solution. In general, therefore, the core is the smaller set. However W. F. Lucas (see "A game with no solution" and "The proof that a game may not have a solution") has shown that a game may have a core but not a Von Neumann-Morgenstern solution. This was a surprising result, because it had been widely conjectured among game theoreticians that every N-person game has at least one Von Neumann-Morgenstern solution, and some effort was made to prove this result as a "fundamental theorem of N-person game theory," analogous to the "fundamental theorem of algebra," according to which every polynomial with complex coefficients has at least one complex root (complex numbers in this context include the real numbers). The hope of providing such a "fundamental theorem" for game theory has now been shattered by Lucas' counter-example. The consolation is that new avenues of investigations have been opened in the mathematics of game theory, namely questions concerning conditions for the existence (or non-existence) of solutions. So stated, the questions may have an interest only for "pure" game theoreticians. However, a social scientist or an economist versed in game theory may be interested in the question of whether such conditions reflect interesting types of social situations formalized as games, as has been found in the case of existence or non-existence of cores (Cf. Chapter 13).

12. The solution consisting of the imputations of the first set is an example of a *symmetric solution.* Each of its imputations can be obtained from another by re-labeling the players. The imputations of the second set are examples of *discriminating solutions.* In each of them a particular player is singled out by being awarded a constant payoff. (Cf. Von Neumann and Morgenstern, *Theory of Games and Economic Behavior,* pp. 288–89.)

13. See W. Vickrey, "Strong and weak solutions in the theory of games" and "Self-policing properties of certain imputation sets."

14. The bargaining set described in this paper is usually designated in the literature as $\mathfrak{M}_1^{(1)}$. Some of its mathematical properties are discussed by M. Davis and M. Maschler ("Existence of stable payoff configurations for cooperative games"). B. Peleg proved that in any such game, there is at least one configuration in $\mathfrak{M}_1^{(1)}$ for every possible coalition structure.

R. J. Aumann and M. Maschler ("The bargaining set for cooperative games") have defined a more general bargaining set $\mathfrak{M}$ in which blocs of players bargain (by raising objections and counterobjections to each other). This set consists of stable *coalitionally* rational payoff configurations. These must satisfy stronger conditions than the stable configurations of $\mathfrak{M}_1^{(1)}$, and consequently there are fewer of them. In particular, in the Divide-the-Dollar game, no stable coalitionally rational configuration corresponds to the grand coalition. Other modified versions of the bargaining set are discussed in that paper.

15. See Davis and Maschler, "The kernel of a cooperative game," and Maschler and Peleg, "A characterization, existence proof and dimension bounds for the kernel of a game; applications to the study of simple games."

16. See Luce and Raiffa, *Games and Decisions,* Chapter 10 and R. D. Luce, "ψ-stability: a new equilibrium concept for n-person game theory."

17. See Chapter 2. This assumption simplifies analysis by relating an individual's prospects to his own "base line," i.e., what he can achieve playing alone against all others.

18. See R. M. Thrall, "N-person games in partition function form."

19. See W. F. Lucas, "On solutions to n-person games in partition function form."

20. There are other methods. See, for example, H. Raiffa, "Arbitration schemes for generalized two-person games" and R. B. Braithwaite, *Theory of Games as a Tool for the Moral Philosopher.* These are discussed also in the author's *Two-person Game Theory,* Chapter 8.

21. Both procedures are based on the same principle, sometimes called the Nash-Zeuthen solution of the bargaining problem. As far as I know, the principle was first derived by F. Zeuthen in 1930 from assumptions concerning the dynamics of bargaining (*Problems of Monopoly and Economic Warfare*). Nash derived the same principle from a set of axioms about the solution (see p. 162). The difference between the Shapley and the Nash procedures derives from different choices of the *status quo* outcome, which obtains if agreement is not reached. I have followed the exposition of Luce and Raiffa (*Games and Decisions,* Chapter 6). Since then I have understood Shapley to say (personal communication) that he is not responsible for what I have called "Shapley's solution" (in *Two-person Game Theory*) and that he believes the Nash solution of the general non-constant-sum two-person cooperative game to be the most acceptable (see J. F. Nash, "Two-person cooperative games"). There may, however, be some justification in the use of Shapley's name in this connection, because, if the payoffs of the two players *can be added* (as is usually assumed in N-person game theory), then what I call here the "Shapley procedure" leads to the Shapley value of the game (see Chapter 5 and Shapley, "A value for n-person games"). Since I understood Shapley to say that he does not object to the use of his name in this connection as an *adjective,* I have taken the liberty to introduce the term "Shapley procedure."

22. See preceding note. The equivalence of Nash's and

Zeuthen's solutions of the bargaining problem was noted by J. C. Harsanyi ("Approaches to the bargaining problem before and after the theory of games: a critical discussion of Zeuthen's, Hick's, and Nash's theories.")

23. Note that the payoffs of a *single* player may be additive even if those of different players are not.

24. M. Shubik (personal communication) has expressed a doubt that Harsanyi's bargaining model leads in all cases to a unique solution (like the Shapley value). I believe that the question still remains open.

25. See J. C. Harsanyi, "A bargaining model for the cooperative n-person game."

26. It might be argued that in case of no sale, players 2 and 3 keep their money, so that $v(\overline{2})$ and $v(\overline{3})$ are not necessarily zero. In view of the normalization of the game, however (cf. Chapter 2), the shift in the "base lines" of the players makes no difference in the analysis of the game.

27. In Figures 10 and 11 (Chapter 4), these triples are represented by the curved lines going from the sides (or vertices) of the inner triangle to the sides of the outer triangle. All such curves are solutions or parts of solutions, provided only they satisfy the conditions mentioned (which imply that the curves must be contained within the three intermediate triangles).

28. See Luce and Raiffa, *Games and Decisions,* p. 233.

29. F. Y. Edgeworth, *Mathematical Psychics.*

30. M. Shubik, "Edgeworth market games."

31. A pure bargaining game is a special case of a weak game, namely one in which every player has a veto.

32. See L. S. Shapley, "Simple games: an outline of the descriptive theory."

33. Note that if a game is not strong, it is not necessarily weak; nor is it necessarily strong if it is not weak. However, it is easily shown that a game cannot be both weak and strong (cf. L. S. Shapley, "Simple games: an outline of the descriptive theory").

34. See R. Bott, "Symmetric solutions to majority games."

35. For further discussion of symmetric games, see B. R. Gelbaum, "Symmetric zero-sum n-person games," L. S. Shapley, "The solutions of a symmetric market game," and D. B. Gillies, "Discriminatory and bargaining solutions to a class of symmetric n-person games"; of quota games, see L. S. Shapley, "Quota solutions of n-person games."

36. See Mann and Shapley, "Values of large games, IV: evaluating the electoral college by Montecarlo techniques."

37. After Luce and Rogow, "A game theoretic analysis of

congressional power distributions for a two-party system." Party 1 is the majority party.

38. See M. Maschler, "The power of a coalition."

39. See W. H. Riker, *The Theory of Political Coalitions.*

40. After Kalisch, Milnor, Nash, and Nering, "Some experimental n-person games."

41. *Ibid.*

42. *Ibid.*

43. After Luce and Raiffa, *Games and Decisions,* Tables 12.3 and 12.4.

44. After M. Maschler, "An experiment on n-person games."

45. See M. Shubik (ed.), *Game Theory and Related Approaches to Human Behavior,* Chapter 24.

46. The rules are reproduced here verbatim (see Note 45). For clarity it is well to keep in mind that "consecutively" here does not necessarily mean "on consecutive moves" but only "in consecutive positions on a pile."

47. Recall, however, that a solution in this sense is a *set* of imputations, not necesarily a single imputation.

48. See, however, note 24.

49. Most of this chapter is based on Gamson's discussion of theories of coalition formation ("Experimental studies of coalition formation").

50. The apparent formation of the "largest" (instead of the minimal) winning coalition (as in nominating conventions) may be due to a bandwagon effect, where everyone joins the winning coalition in order "to be on the winning side" or to avoid political reprisals.

51. After Vinacke and Arkoff, "An experimental study of coalitions in the triad."

52. See B. Lieberman, "Experimental studies of conflict in some two-person and three-person games."

53. See W. H. Riker, *The Theory of Political Coalitions.*

54. One context in which this fluidity in the identity of the actor is sometimes taken into account is where shifting coalitions are at the focus of attention of a strategist. Nevertheless an *individual* set of interests (of a firm, a nation, or the like) is always foremost in these deliberations. This is the "melody"; the rest is "accompaniment." Not so in game theory, where all possible coalitions demand equal attention simultaneously ("in counterpoint"). In this connection, see Chapter 11.

55. The national state (let alone its potentate or elite) is no longer universally regarded as the only legitimate actor in international relations, something taken for granted less than a century ago.

References

Aumann, R. J. and Maschler, M. "The bargaining set for cooperative games," in Dresher, M., Shapley, L. S., and Tucker, A. W. (eds), *Advances in Game Theory* (Annals of Mathematics Studies, 52), Princeton, N.J.: Princeton University Press, 1964.

Bott, R. "Symmetric solutions to majority games," in Kuhn, H. W. and Tucker, A. W. (eds), *Contributions to the Theory of Games,* II (Annals of Mathematics Studies, 28), Princeton, N.J.: Princeton University Press, 1953.

Braithwaite, R. B. *Theory of Games as a Tool for the Moral Philosopher.* Cambridge: Cambridge University Press, 1955.

Davis, M. and Maschler, M. "Existence of stable payoff configurations for cooperative games." *Bulletin of the American Mathematical Society,* 69 (1963), 106–8.

Davis, M. and Maschler, M. "The kernel of a cooperative game." *Naval Research Logistics Quarterly,* 12 (1965), 223–59.

Edgeworth, F. Y. *Mathematical Psychics.* London: Kegan Paul, 1881.

Gamson, W. "Experimental studies of coalition formation," in L. Berkowitz, (ed.), *Advances in Experimental Social Psychology,* I, New York: Academic Press, 1964.

Gelbaum, B. R. "Symmetric zero-sum n-person games," in Luce, R. D. and Tucker, A. W. (eds), *Contributions to the Theory of Games,* IV (Annals of Mathematics Studies, 40), Princeton, N.J.: Princeton University Press, 1959.

Gillies, D. B. "Discriminatory and bargaining solutions to a class of symmetric n-person games," in Kuhn, H. W. and Tucker, A. W. (eds), *Contributions to the Theory of Games,* II (Annals of Mathematics Studies, 28), Princeton, N.J.: Princeton University Press, 1953.

Harsanyi, J. C. "Approaches to the bargaining problem before and after the theory of games: a critical discussion of Zeuthen's, Hick's, and Nash's theories." *Econometrica,* 24 (1956), 144–57.

———. "A bargaining model for the cooperative n-person game," in Luce, R. D. and Tucker, A. W. (eds), *Contributions to the Theory of Games,* IV (Annals of Mathematics Studies, 40), Princeton, N.J.: Princeton University Press, 1959.

Kalisch, G. K., Milnor, J. W., Nash, J. F., and Nering, E. D.

"Some experimental n-person games," in Thrall, R. M., Coombs, C. H., and Davis, R. L. (eds), *Decision Processes*. New York: John Wiley & Sons, 1954.

Kannai, Y. "Continuity properties of the core of a market game." Research Program in Game Theory and Mathematical Economics, Research Memorandum 34, January 1968. Jerusalem: Hebrew University of Jerusalem (mimeo).

Lieberman, B. "Experimental studies of conflict in some two-person and three-person games," in Criswell, J. H., Solomon, H., and Suppes, P. (eds), *Mathematical Methods in Small Group Processes*. Stanford: Stanford University Press, 1962.

Lucas, W. F. "On solutions to n-person games in partition function form." Ph.D. thesis at The University of Michigan, Ann Arbor, 1963.

———. "A game with no solution." Memorandum RM-5518-PR, November 1967. Santa Monica: The Rand Corporation.

———. "The proof that a game may not have a solution." Memorandum RM-5543-PR, January 1968. Santa Monica: The Rand Corporation.

Luce, R. D. "ψ-stability: a new equilibrium concept for n-person game theory." *Mathematical Models of Human Behavior*. Stamford, Conn.: Dunlap and Associates, 1955.

Luce, R. D. and Raiffa, H. *Games and Decisions*. New York: John Wiley & Sons, 1957.

Luce, R. D. and Rogow, A. A. "A game theoretic analysis of congressional power distributions for a two-party system." *Behavioral Science*, 1 (1956), 83–95.

Mann, I. and Shapley, L. S. "Values of large games, IV: evaluating the electoral college by Montecarlo techniques." Memorandum RM-2651, September 1960. Santa Monica: The Rand Corporation.

Maschler, M. "An experiment on n-person games." *Recent Advances of Game Theory*. Princeton University Conference, 1962.

———. "The power of a coalition." *Management Science*, 10 (1963), 8–29.

Maschler, M. and Peleg, B. "A characterization, existence proof and dimension bounds for the kernel of a game; applications to the study of simple games." *Pacific Journal of Mathematics*, 18 (1966), 289–328.

Nash, J. F. "Two-person cooperative games." *Econometrica*, 21 (1953), 128–40.

Peleg, B. "Existence theorem for the bargaining set $\mathcal{M}_1^{(i)}$." *Bulletin of the American Mathematical Society*, 69 (1963), 109–10.

Raiffa, H. "Arbitration schemes for generalized two-person games," in Kuhn, H. W. and Tucker, A. W. (eds), *Contributions to the Theory of Games,* II (Annals of Mathematics Studies, 28), Princeton, N.J.: Princeton University Press, 1953.

Rapoport, A. *Two-person Game Theory: The Essential Ideas.* Ann Arbor: The University of Michigan Press, 1966.

Riker, W. H. *The Theory of Political Coalitions.* New Haven: Yale University Press, 1962.

Shapley, L. S. "A value for n-person games," in Kuhn, H. W. and Tucker, A. W. (eds), *Contributions to the Theory of Games,* II (Annals of Mathematics Studies, 28), Princeton, N.J.: Princeton University Press, 1953.

———. "Quota solutions of n-person games," in Kuhn, H. W. and Tucker, A. W. (eds), *Contributions to the Theory of Games,* II (Annals of Mathematics Studies, 28), Princeton, N.J.: Princeton University Press, 1953.

———. "The solutions of a symmetric market game," in Luce, R. D. and Tucker, A. W. (eds), *Contributions to the Theory of Games,* IV (Annals of Mathematics Studies, 40), Princeton, N.J.: Princeton University Press, 1959.

———. "Simple games: an outline of the descriptive theory." *Behavioral Science,* 7 (1962), 59–66.

Shubik, M. "Edgeworth market games," in Luce, R. D. and Tucker, A. W. (eds), *Contributions to the Theory of Games,* IV (Annals of Mathematics Studies, 40), Princeton, N.J.: Princeton University Press, 1959.

———. (ed.). *Game Theory and Related Approaches to Human Behavior.* New York: John Wiley & Sons, 1964.

Thompson, G. L. "Signaling strategies in n-person games," in Kuhn, H. W. and Tucker, A. W. (eds), *Contributions to the Theory of Games,* II (Annals of Mathematics Studies, 28), Princeton, N.J.: Princeton University Press, 1953.

Thrall, R. M. and Lucas, W. F. "N-person games in partition function form." *Naval Research Logistics Quarterly,* 10 (1963), 281–98.

Vickrey, W. "Strong and weak solutions in the theory of games." Department of Economics, Columbia University, 1953 (dittoed).

———. "Self-policing properties of certain imputation sets," in Luce, R. D. and Tucker, A. W. (eds), *Contributions to the Theory of Games,* IV (Annals of Mathematics Studies, 40), Princeton, N.J.: Princeton University Press, 1959.

Vinacke, W. E. and Arkoff, A. "An experimental study of coalitions in the triad." *American Sociological Review,* 22 (1957), 406–15.

Von Neumann, J. and Morgenstern, O. *Theory of Games and Economic Behavior,* 2nd ed. Princeton, N.J.: Princeton University Press, 1947.

Zeuthen, F. *Problems of Monopoly and Economic Warfare.* London: G. Routledge and Sons, 1930.

Index

THE EVOLUTION OF MAN *by G. H. R. von Koenigswald*

SPACE CHEMISTRY *by Paul W. Merrill*

THE LANGUAGE OF MATHEMATICS *by M. Evans Munroe*

CRUSTACEANS *by Waldo L. Schmitt*

TWO-PERSON GAME THEORY: THE ESSENTIAL IDEAS *by Anatol Rapoport*

THE ANTS *by Wilhelm Goetsch*

THE STARS *by W. Kruse and W. Dieckvoss*

LIGHT: VISIBLE AND INVISIBLE *by Eduard Ruechardt*

THE SENSES *by Wolfgang von Buddenbrock*

EBB AND FLOW: THE TIDES OF EARTH, AIR, AND WATER *by Albert Defant*

THE BIRDS *by Oskar and Katharina Heinroth*

PLANET EARTH *by Karl Stumpff*

ANIMAL CAMOUFLAGE *by Adolf Portmann*

THE SUN *by Karl Kiepenheuer*

VIRUS *by Wolfhard Weidel*

SOIL ANIMALS *by Friedrich Schaller*

N-PERSON GAME THEORY: CONCEPTS AND APPLICATIONS *by Anatol Rapoport*